Man of Tempered Steel - Bruno Schlesinger

*Biography of my Father: An Early
South African Mining Engineer*

By

Helga Kaye with **Keith W Kaye**

Edited and with Contributions by:

Keith Woodhill Kaye. BSc(Hons), MB.Bch. FRCS(Edin), FCS(South Africa), FRACS.
American Boards in Urology

AuthorHouse™
1663 Liberty Drive,
Suite 200
Bloomington, IN 47403
www.authorhouse.com
Phone: 1-800-839-8640

AuthorHouse™ UK Ltd.
500 Avebury Boulevard
Central Milton Keynes, MK9 2BE
www.authorhouse.co.uk
Phone: 08001974150

©2007 Helga Kaye with Keith W Kaye. All rights reserved.

No part of this book may be reproduced, stored in a retrieval system, or transmitted by any means without the written permission of the author.

First published by AuthorHouse 2/6/2007

ISBN: 978-1-4259-7963-8 (sc)

Library of Congress Control Number: 2006910760

Printed in the United States of America
Bloomington, Indiana

This book is printed on acid-free paper.

Bloomington, IN   Milton Keynes, UK
authorHOUSE®

Figure 1.
Bruno about 1915

DEDICATION

To those who went before and guided me on the path, especially my parents, Josse and Helga Kaye. To those who enjoy the journey and adventure and make it so worthwhile especially my wife Valda, children Jessica, Deborah and Maxine and my sister Rea Gardy. To all my widespread, varied, interesting family who touch my life in so many wonderful ways. To those who will come after so they may have some glimpses of their heritage.

Keith W Kaye
Minneapolis. November 2006

CONTENTS

Introduction and Acknowledgements	vii
Chapter 1. First Memories. 1912	1
Chapter 2. Origins, Parents, Coming to Africa, Marriage. 1750's – 1907	3
Chapter 3. Early Times as Mining Engineer in God's Own Country. 1907-1910	23
Chapter 4. World War 1, Anti-German Sentiment, Internment, Mozambique. 1914-1917	32
Chapter 5. Barman, Manager Diamond Mine and Box Factory. Return to Profession. 1917 - 1922	34
Chapter 6. Life as Mining Engineer. 1924 - 1934	39
Chapter 7. Family. A Good Friend. 1935 - 1942	55
Chapter 8. Illness and Heart Attack. 1943-1945	63
Appendix 1. Descendant Tree of Isidor Schlesinger	67
Appendix 2. Hourglass Tree of Bruno Schlesinger	68
Appendix 3. Descendant Tree of Helga Schlesinger (Kaye)	69
Appendix 4. Descendant Tree of Rolf Schlesinger	70
Appendix 5. Descendant Tree of Feodor Schlesinger	71
Appendix 6. Descendant Tree of Valerie Schlesinger (Pollak)	72
Appendix 7. Law Suit Isidore Schlessinger: 1866	74
Appendix 8. Law Suit: Schlesinger v Donaldson and Carlis 1928	76

Figure 2.
Preparing book at "Woodhill Sanctuary", overlooking Roaring Creek, Belmopan, Belize

INTRODUCTION AND ACKNOWLEDGEMENTS

This book was born as an essay that my mother, Helga Kaye (nee Schlesinger), wrote a few years before her death in 1998. It hibernated until several ago, when I became interested in our family history. At that stage I roughly divided the essay into chapters and made some additions. Since then a significantly greater amount of information and number of documents and photographs have to come light.

During one of my regular visits to Belize, where I do voluntary urology work at the Western Regional Hospital in Belmopan, and have started a teak plantation, I took off some time to get this project underway. This week was a 'dream time'. For several days I worked quietly at our condominium in Placencia, which is lapped by the beautiful island-studded Caribbean Sea. Afterwards I returned to our property, "Woodhill Sanctuary", near Belmopan, and continued the delightful task. Sitting within the insect free environment of our screened porch I listened to innumerable bird sounds; brown jays, woodpeckers, toucanets and even the screech of parrots. This contrasted with the restful song of Roaring Creek, which could be seen glistening through the dense foliage below our jungle cottage. Here I could think about my family. I was in paradise!

I would like to thank the following for their help in so many ways: My sister and niece, Rea and Tanya Gardy for reviewing the text and excellent suggestions, especially regarding the cover; cousin Jenny Fasal for her tremendous work in digging out the Fasal Family History, a superb family reunion, and permission to use Fasal documents; cousin Erica Wolter for the fun in exploring our ancestors together and permission to use old family paintings; Markus and Sabina Gapany, Simon and Esther Birnbaum, Gerta Birnbaum, Waclaw Jaszcz, Alexandra Helm, Ursel and Herbert Goldschmidt for their translations; Bruce Larson - lawyer, friend and neighbour for interpreting the old lawsuits; Jeanne and Scott Hollington for the superb maps; and as always, my wife Valda, and children Jessica, Deborah and Maxine for their love, support and encouragement.

Keith W Kaye

Minneapolis November 2006

Figure 3.
Kroonstad, Christmas 1911; L-R: Bruno, Mrs. Hooper, Else, Helga, Mr. Hooper.

CHAPTER 1.

First Memories.
1912

One of my earliest recollections is of my father dragging my brother and me, first to the front gate, then to the back gate, of the little mine house we occupied in Maraisburg, South Africa.

"Don't either of you ever dare set foot out of these gates again without your mother's permission," he said sternly.

Then we were each given a thrashing and sent to bed.

I was five years old in 1912, my brother three. We had been playing 'house-house', when I suddenly decided it was time we saw something of the world! I took Rolf by the hand and walked down the country road towards the mine where our father was underground manager.

Skirting the headgear we wandered through a wood and when we emerged on the other side, I recognized a house that we had visited with our parents. I remembered especially the sweets and delicious cake Mrs. Hooper had given us. Gripping my little brother's hand more firmly I ran across the road and knocked at the door. Mrs. Hooper's expression of surprise that we had come all this way alone made me feel very important and independent. I accepted with alacrity her kind offer of some refreshment before she took us back home. Neither she nor our parents had a telephone.

We were enjoying our milk and biscuits, when hurried steps were heard on the verandah and the door bell harshly rang. Our mother's agitated cry still rings in my ears:

"Mrs. Hooper I have lost my children!"

Our faithful Matabele servant, Basumba, who remained with my parents until the outbreak of World War 1, when we had to leave the mine, had followed our footsteps in the dusty earth. When these led to the wood, Mother was almost frantic. Several murders had recently been committed there. We however, had experienced nothing more frightening than a dog which barked at us and sent us scurrying faster along the path.

Our father's anger and the thrashing we received were far more frightening than the encounter with the dog.

Map 1. Central Europe with today's borders showing cities
and towns lived in or visited by Schlesinger's and Fasals.

CHAPTER 2.

Origins, Parents, Coming to Africa, Marriage.
1750's – 1907

The name Schlesinger originates from the region of Central Europe known as Silesia (Schlesien in German) and is a well known Jewish name. Silesia now, is mainly within South Western Poland with small parts in North Eastern Czech Republic and in Germany. Major cities are Wroclaw and Katowice. The area however, has had a long and varied history, with a strong Germanic presence. In 1526, with the Holy Roman Empire, it passed to the crown of the Habsburg Monarchy of Austria. This seems to have been the start of the Schlesinger love affair with Austria. After the War of Austrian Succession (1740-1748) the much loved Queen Maria Theresa had to cede most of Silesia to Prussia (King Frederick the Second). The Duchies of Teschen (Cieszyn) and Troppau (Opava) however remained within the Habsburg Dynasty, were known as Austrian Silesia, and were part of the Austrian Empire from 1804 to 1867 and the Austro-Hungarian Empire from 1867 to 1918. This explains the strong feelings the Schlesinger's had for Austria and especially its glorious capital city of Vienna.

Bruno's father Isidor was born on 10 March 1842. Exactly where he was born we are not sure. From the Degistration Certificate (Figure 12) it states Kempeny, however we cannot find where that is. Interestingly this document gives his year of birth as 1843. The Government Regional President letter, Opole, 31st August 1891, maintains he was born in the Province of Posen (Figures 14 and 15).

Bruno's mother, Emma Fasal was born in Wyrubana number 69, which is part of the village of Chotebuz, about 4 km from Teschen (now Cesky Tesin in Czech Republic) on 22 July 1854. The word Wyrubana means 'the place where the forest was sawn down ages ago'. There is some discrepancy as to whether she was born in June or July 1854. An 1885 copy of her birth certificate shows July as do most other documents, whereas some papers state 22.6.1854.

As they were growing up, initially in the Austrian and then the Austro-Hungarian Empires, Teschen was under the nominal rule of the Habsburg, Archduke Albert, Duke of Teschen (1817-1895). Later, between World War 1 and 2, the city was divided with the West side of the Olza River becoming Cesky Tesin within Czechoslovakia, and the Eastern side Cieszyn in Poland. This explains why Emma's 1933 passport was issued by the then country of Czechoslovakia.

Isidor was a very adventurous young man and before the age of 24 in 1866, had already left the security of Europe to make his fame and fortune in South Africa; which indeed he did. Initially, traveling by ox-wagon, he crossed the veld to Pilgrim's Rest in the Eastern Transvaal, which was the Eldorado of those distant times. Later he went to the Kimberly Diamond Mines. Besides the undeniable adventures and excitement of a new land, new peoples, gold and diamonds, this also seems to have been a most litigious time. There is a Cape Supreme Court file of 30 handwritten pages dealing with lengthy and involved cases in which firstly Isidore Schlessinger (as it was spelled at that time) of Cape Town is suing Richard Rutherford for amounts collected by Rutherford but not paid over to Schlessinger. Then there are cases in which Schlessinger in turn is being sued for amounts he owed to other persons and firms. There are several such claims and counterclaims and much evidence from various witnesses. (Appendix 7).

When Isidor (the spelling of both first and surnames seems to have changed about now) eventually returned to Europe some 7-8 years later he was a rich man and must have cut a dashing figure. At the age of 30 he met, and fell in love, with the tall, elegant Emma Fasal. As can be seen from his wedding certificate, at the time of his marriage in 1874 he was living in Bielsko (now in Polish Silesia), whereas Emma still lived in the town of her birth, Teschen (now Cesky-Tesin in the Czech Republic).

Figure 4. Teschen (Cieszyn) in the time of the Austrian Empire with the River Olza

Figure 5. From Moritz Fasals' Liquor Factory with Imperial Eagle.

Emma was from a well to do Jewish Austrian family. All photographs show her as a very tall, imperious looking woman, inevitably wearing a neckerchief. Her family can be traced back to Salomon Lobl, a so called 'Tolerated Jew' who under Maria Theresa's Edict of Tolerance of 1752 was officially permitted to live in the Cieszyn region. Salomon also lived in Chotebuz (Polish Kocobedz) and thus we know that the family lived in this small village for at least 100 years. At the present time Chotebuz has a population of around 1000 individuals. For some reason Salomon began to sign himself as Salomon Fasal. Emma's father, Josef Fasal, a descendant of Salomon, was a merchant and owner of a house on the left bank of the Olza River in the quarter known as Kamieniec at 56 Saska Kepa. Josef married Charlotte (Lotte) Neumann. They had three children Ferdinand (born 1842), Fanny (born 2 September 1848) and the youngest Emma who seems to have been the last Fasal born in Chotebuz. Interestingly Fanny married her cousin Maurycy (Moritz) Fasal (1841-1919) who also lived in the house at 56 Saska Kepa. A tavern functioned at this address which became famous in stories in the Cieszyn Star from 1861 - 1873. Moritz in 1868 went on to found the well known M. Fasal Liquor Company which won numerous International prizes and awards among others in Brussels in 1892 and in London in 1893. The company was also given the great honour to use the Imperial Eagle on its labels.

Bruno's younger sister Leontine, (who was later to become the famous actress and producer Leontine Sagan) wrote in her published autobiography 'Lights and Shadows'; "the Fasals were well-to-do respectable people, who had completely assimilated with the lively, cheerful middle class of old Austria. Their children went to schools and universities and in no way differed from other young Austrians. Indeed so great was their attachment to their home country that they did not even want to spend their holidays abroad, because they could not bear to be separated from Austrian scenery, Austrian customs and Austrian cookery."

Emma, like her husband, was adventurous in spirit, and indeed was called "our adventuress" by her family. By September 1877 Emma and Isidor had two children, Valeria and Feodor. At this time they moved from Kattowitz (now Katowice in Poland) to Troppau (now Opava, Czech Republic) which at that time was the capital of Austrian Silesia and a part of the Austro-Hungarian Empire. Here they set up a saw-mill and Bruno was born in Troppau on the 22nd March 1879.

Bruno was thus born into, and also imbued with, Austrian Culture. This seems to have percolated down through the generations. Keith well remembers his mother Helga, proudly saying she was Austrian (although she was born in South Africa) and that Vienna was the City of her heart. When Bruno's niece Dr Ilse Wolf (Schlesinger) died she wished her ashes to be in strewn in Vienna and indeed her daughter Erica Wolter fulfilled these wishes by laying them to rest at the foot of the monument to Johann Strauss.

Unfortunately Isidor's saw-mill was not a success. Though he was a kindly, gentle and strictly honest man, he was completely impractical, and soon lost the fortune he had so diligently accumulated in South Africa. For some time after Bruno was born, Emma and Isidor roamed around to different places, until finally in Budapest, shortly after their daughter Leontine was born in 1890, the money ran out. Having tasted the wilds of Africa, Isidor never really felt happy in Europe and the call of the wide open spaces lured him back to the Dark Continent. He felt certain he would retrieve his wealth and independence on the Witwatersrand goldfields and could then send for his wife and children. It would be eight years however, before they were able to join him.

Emma was then left to fend for herself and her four children, ranging in age from one year old Leontine to fourteen year old Valerie. After Isidors' departure she moved to Vienna, the spiritual home of the Fasals and Schlesinger's and probably the city of her mother's family, the Neumann's,

Emma had certainly been nurtured in all the luxury of a refined, cultured home. She was however, a dynamic, resourceful woman of above average intelligence. With the change in her fortunes and her husband leaving to try to redeem them in South Africa, Emma set about earning a living for her children and herself. She was determined to give them a good education and provide a happy home. After many vicissitudes she obtained a position with the Austrian State Dairy. This necessitated her traveling to farms in all the outlying districts to purchase milk, butter, cream and cheese from the peasants.

Figure 6. This painting dated 1836 is almost certainly that of Bruno's grandmother Charlotte (Lotte) Neumann. It was given by Bruno's sister Leontine Sagan to her niece Dr. Ilse Schlesinger (Wolf).

Figure 7. Painting in original frame with exact necklace being worn by young girl.

Leontine, further describes this time; "Mother immediately set about earning our livelihood. In this difficult task she was guided more by her energy and curiosity than by the knowledge of business; and the still young and beautiful woman flung herself into a struggle to which she was unsuited. As far back as I can remember, financial worries overshadowed her joyous, optimistic nature but she was determined to give her children a good education and to establish for herself and for us a happy hospitable home. Tired out after a strenuous day, she would quickly change into an evening dress and escort my sister to a dance or theatre, she herself being passionately fond of both. Her lively temperament never allowed her to be depressed or discouraged, no matter how tedious or ludicrous her business experiences might be. These varied according to her enterprises and experiments. She would try anything as long as it didn't entail sitting around in an office." During her time with the Austrian State Dairy she entangled "herself in all sorts of absurd affairs. A paunchy mayor would try to make love to her, or the innkeeper's wife would honour her specially by offering her the bed of her husband who had died the night before. Mother used to have her family and friends in fits of laughter with her vivid descriptions of Austrian peasantry which differed widely from the idealized version of the country's novelists and poets."

Meanwhile, Vali, the eldest child, was looking after the household and especially little Leontine. The two boys, Feodor and Bruno were still at school.

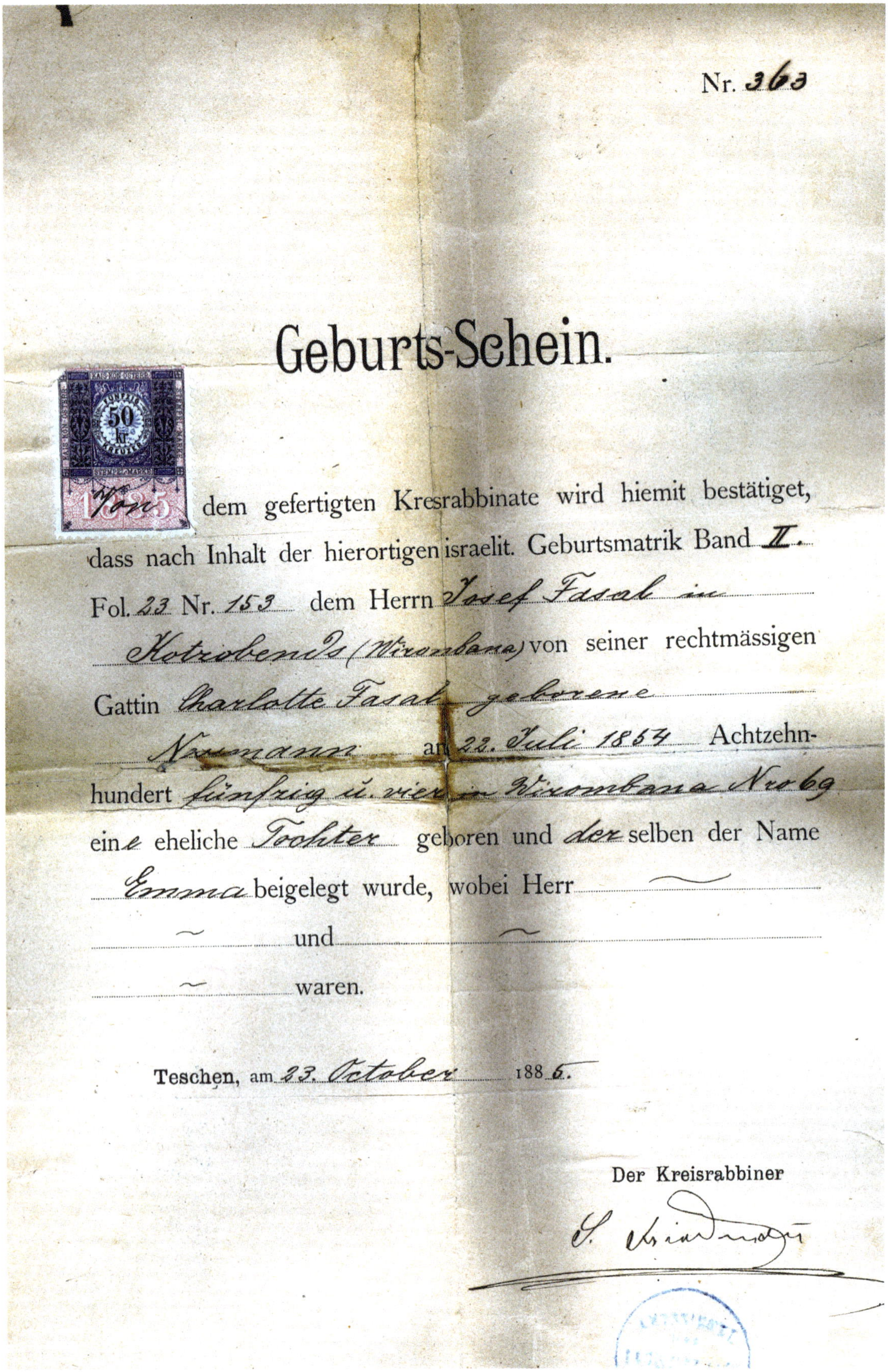

Nr. 363

Geburts-Schein.

Von dem gefertigten Kresrabbinate wird hiemit bestätiget, dass nach Inhalt der hierortigen israelit. Geburtsmatrik Band II. Fol. 23 Nr. 153 dem Herrn Josef Fasal in Kotzobendz (Wirombana) von seiner rechtmässigen Gattin Charlotte Fasal geborene Neumann am 22. Juli 1854 Achtzehnhundert fünfzig u. vier in Wirombana Nro 69 eine eheliche Tochter geboren und der selben der Name Emma beigelegt wurde, wobei Herr ~ ~ und ~ ~ waren.

Teschen, am 23. October 188 5.

Der Kreisrabbiner

S. Friedmann

Figure 8. Copy of Emma's 1854 Birth Certificate – Issued in Teschen in 1885.
Note; *"in Wirombana Nro 69"*.

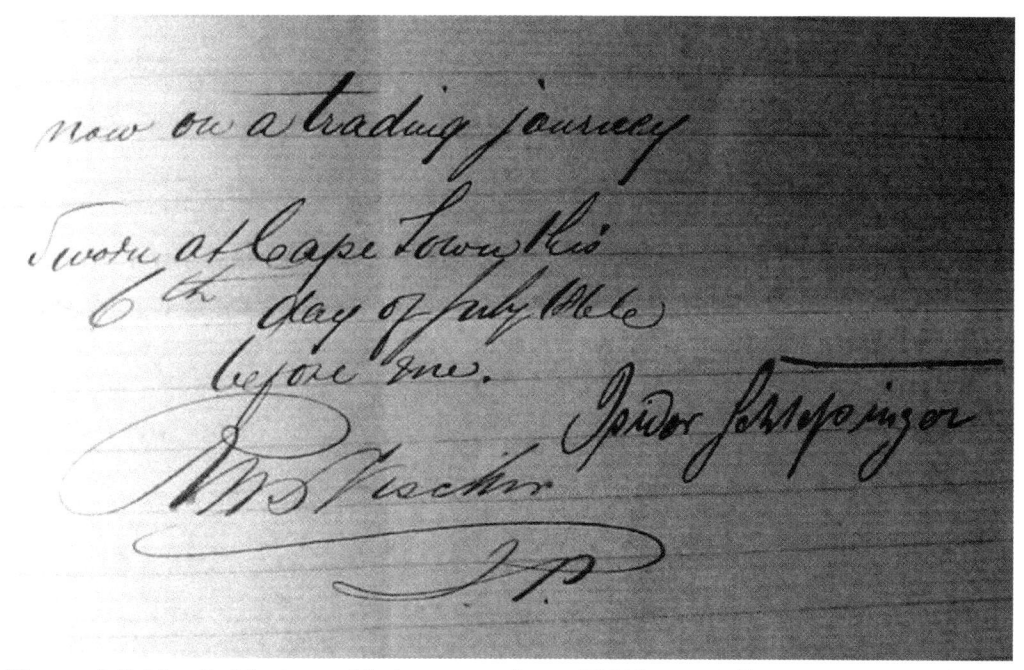

Figure 9. Isidor Schlesinger's' signature from 1866 lawsuit "… now on a trading journey. Sworn at Cape Town this 6th Day of July 1866 before me …"

At last, in 1899, the family set sail for South Africa, leaving Bruno behind to complete his studies at the, to this day, famous School of Mines in Leoben, Austria. On arriving in South Africa, Leontine, age 9, first got to know her father and later describes him in these three vignettes:

Isidor 1:
"When we arrived in Cape Town (1899) a letter from my father was brought on board, in which he told us that the Transvaal was on the eve of war and he dare not leave his business in Klerksdorp, but would meet us in Kroonstad, in the Orange River Colony,' the terminus of the railway journey. Thus the beginning of our new life was overshadowed with anxious forebodings.

In the late afternoon of a South African winter's day our train arrived at a little corrugated iron house in the veld which was Kroonstad's railway station. My family's excitement, the Babel of foreign languages - English, Dutch and Native - is stamped in my mind. Out of the darkness I heard a man calling our names. `How does this strange man know our names?' passed through my mind. I felt almost like a spectator, I, the only one who did not know my father. Even in later years I could never overcome that strangeness. For eight years he had not seen his wife and children. Mother was then forty-five years old, still a strikingly beautiful woman, and it seemed curious to me that a `stranger' should embrace her and that my sister and brother should call him Papa."

Isidor 2:
"There lived in Klerksdorp only a few European families; English, Dutch and German. The little town was divided into an old and a new section. The old part lay across the dry river bed; it had been founded by the Voortrekkers and consisted only of a few farms and a church. In the new town my father owned the bar at the Freemason Lodge. The many stages of his pioneering had taken him no further on the road to success. One could not have imagined a man less suited to his job. He was a dreamer by nature, cared little about wealth, and felt happiest when he could sit with his pipe by the open veld-fire or with a book on the stoep. His friends included Afrikaners, Englishmen, and a few Germans, who had lived in the country for many years and who shared both his love for South Africa and his indifference to Europe. Their conversations circled around their business, the share-market in Johannesburg, politics, and that soft, gentle gossip which is a feature of every small town."

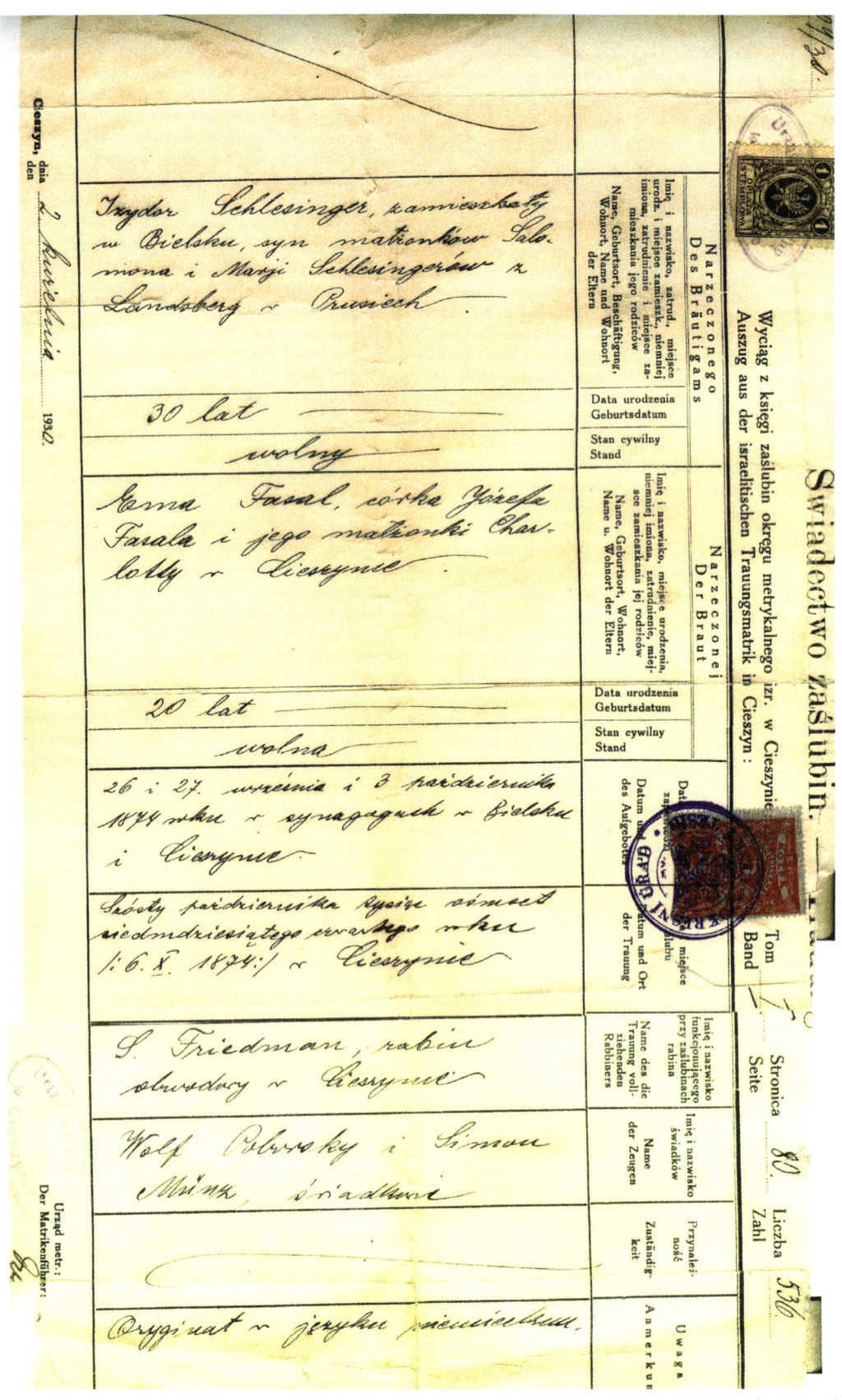

Figure 10. 1930 Copy of Emma and Isidor's 1874 Wedding Certificate. Written in Polish and issued in Cieszyn, Poland. There are two stamps. One from Poland and the other from Czechoslovakia. This document was probably obtained by Emma when she was planning to come and live in South Africa permanently. In 1935 she obtained a Czechoslovakian passport and was a resident of Cesky/Tesin

Marriage Certificate

This is an excerpt from the records of marriages in the Jewish Register in Cieszyn City. Vol. 1 Page 80. (A copy of the original issued in Cieszyn on 2 April 1930.)

Bridegroom:

Isydor Schlesinger, residing in Bielsku, son of Salomona and Marji Schlesingerow from Landsberg in Prussia.

Age: *30 years*

Legal status: *Not married*

Bride:

Ema Fasal, daughter of Josefa Fasala and Charlotty from Cieszyn

Age: *20 years*

Legal status: *Not married*

Date and Place of Banns:

26 and 27 September and 3 October, 1874 in synagogues in Bielsku and Cieszyn.

Date and Place of Marriage:

October 6, eighteen hundred seventy four (6.10.1874) in Cieszyn

Name of Officiating Rabbi:

S. Friedman, District Rabbi of Cieszyn

Witnesses: *Wolf Pobrosky and Simon Munz*

Comment: *Original document in German language*

Figure 11. Translation of Copy of Wedding Document (Translated by Dr Waclaw Jaszc)

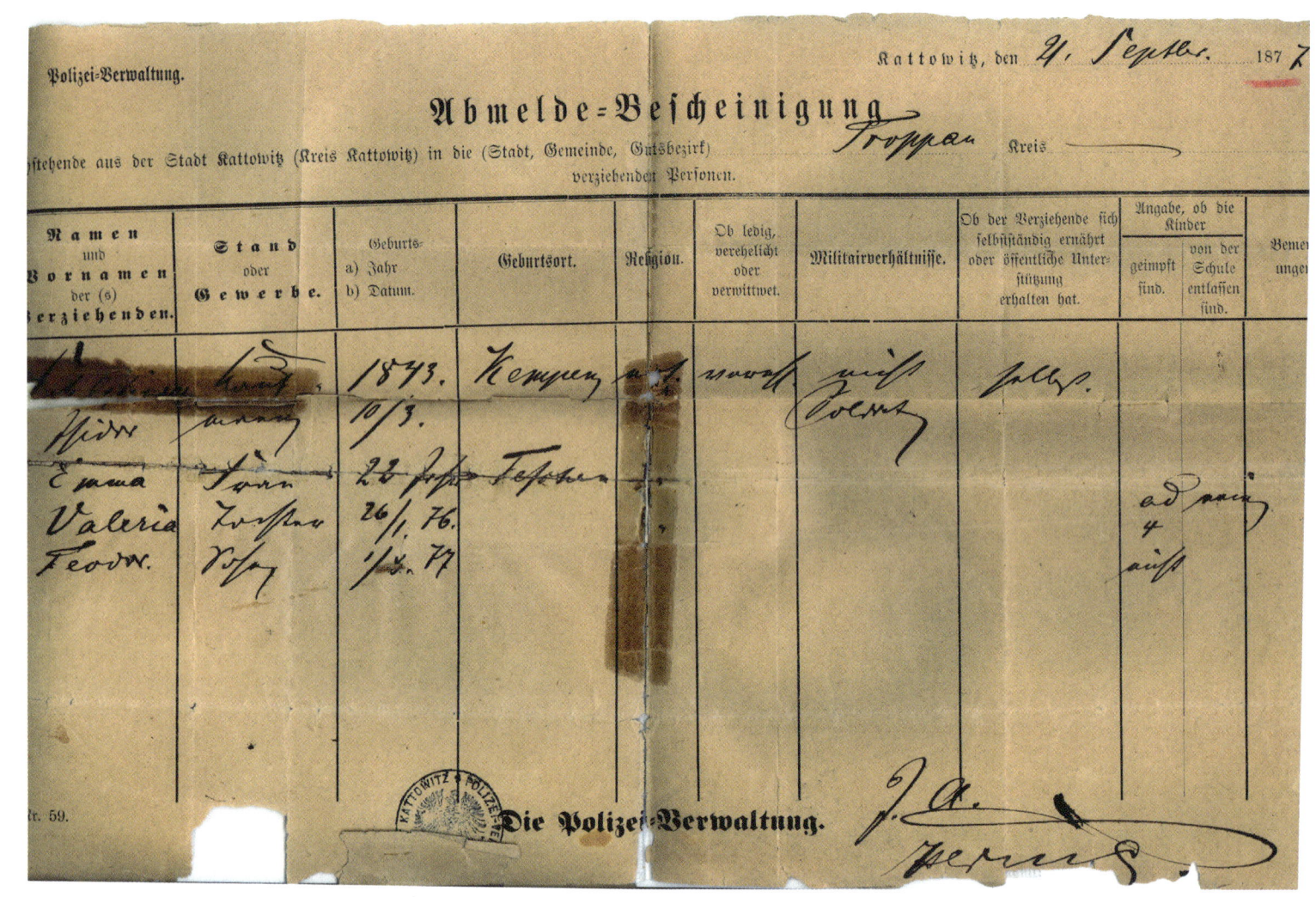

Figure 12. Deregistration Certificate. Issued by Police Registry of Kattowitz. *21 September 1877.* For the people named (*Isidor Schlesinger* born in *Kempeny*, married, *Emma* born in *Teschen*, *Valeria*, *Feodor*) who are moving to *Troppau*. Can the people moving look after themselves or are they on Welfare – *self sufficient*. Vaccination, Schooling. This was before Bruno was born and suggests that Valeria (18mths old) and Feodor (one month old) were probably born in Kattowitz.

Isidor 3:

"He had a good word to say about everyone and would not tolerate snobbery. South Africa, he would say, was a blessed country which knew no class distinctions and where character alone counted. (It has changed considerably in modern times.) Had not he himself descended from his former social status and did anyone respect him less? His friends were kind to his wife and children and ungrudgingly admired their fine education and worldly manners, so why disparage them? All the same, Klerksdorp WAS different from Vienna!"

Writing about the 'uitlanders' or foreigners, in South Africa at this time during the Boer War Leontine continues; "Austrians and Germans were not considered enemies as we had complete freedom to come and go as we pleased. For the neutral 'uitlander', the Boer War could be called almost a comfortable one. The battle field was far away, food stuffs were rationed, but also bartered and exchanged between friends; bags of flour and sugar, butter and eggs being welcome birthday and wedding gifts. A small colony of Austrians and Germans settled down to intimate companionship. …Dances, picnics and concerts prevailed."

In South Africa, the German and Austrian communities naturally drew together.

Figure 13. 1920 Copy of original Bruno Birth Certificate. Translated as: **BIRTH CERTIFICATE** *no 65*. The Jewish registry herewith acknowledges that according to the content of the Jewish Birth Registry Vol *11*, Page *14*, entry *183*, Mr. *Isidor Schlesinger, mill owner* in *Troppau, Kemsrauerg 23,* and his wedded wife *Emma, nee Fasal,* on 22. *March, 1879* in *Troppau*, a legitimate *son* was born, and was given the name *Bruno*. Troppau, *16. June* 1920. Jewish Registry. *Schlesinger. Substitute registrar.*

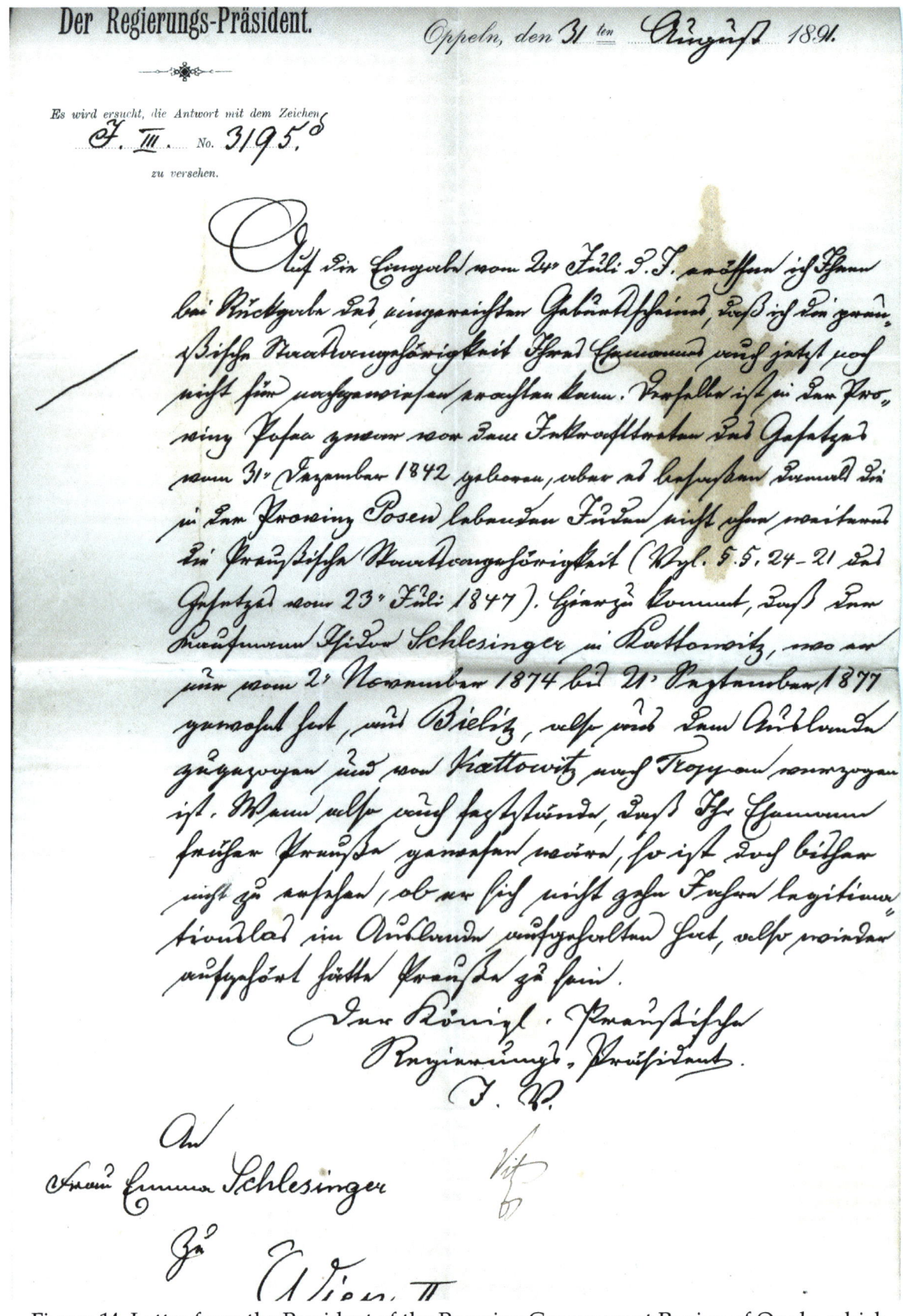

Figure 14. Letter from the President of the Prussian Government Region of Opole, which was the capital of Upper Silesia, to Emma Schlesinger, denying Prussian Citizenship to her husband, Isidor Schlesinger. 31st August 1891.

The Government Regional President

Opole 31 st August 1891

In response to you enquiry re :
F III. No. 3195 d

From your request of 24 th July of this year I am returning your husband's birth certificate however I still cannot acknowledge the Prussian Citizenship of your husband. Your husband was born in the Province of Posen before the law was passed on December 31 st 1842. At that time not every Jew living in the Province of Posen could normally obtain Prussian Citizenship. (see paragraph 24321 from the Law of 23 rd July 1847) Another issue is that the merchant, Isidor Schlesinger, only lived in Kattowitz (Katowice) from 2 nd November 1874 until 21 st September 1877. He moved there from Bielsko (Bielitz) which was foreign territory and then moved from Kattowitz (Katowice) to Troppau (Opava). Even if your husband was a Prussian Citizen before, it is not clear that he did not lose his Citizenship by legitimately living outside of Prussia for ten years.

The Royal Prussian
Government Regional President

To

Frau Emma Schlesinger

In

Vienna II

Figure 15. Translation of Figure 14.

The family of Heinrich and Helene Gimkewitz with their children, Else (born in Berlin on 18th August 1882), Friedel, Sophie and Fritz and that of Isidor and Emma Schlesinger, with their children Vali, Feodor, and Leontine soon established a close friendship. Else was the eldest of the Gimkewitz girls and when she came to South Africa in 1899 she had just finished a course of study at the Karl University in Prague. As the family friendship developed, Else began writing to the young Austrian student still completing his studies in Europe. Bruno arrived in South Africa in 1900 and a whirlwind courtship ensued. Nonetheless, the parents on both sides were against an early marriage.

Bruno's mother Emma, who had temporarily returned to Europe with Leontine, wrote from Prague on 9 September 1903:

"My Dearest Bruno,
It is now the second time that I have your destiny in my hands. When I dissuaded you from choosing a profession that would never have contented you, you alone were concerned. In this case however, two of you are concerned – you and your – our - Else. I am happy to give you my blessing, because I could not wish for a better wife for you.

And now because I love Else and have the greatest regard for her parents, I must warn you to be prudent, my beloved child. When I return we shall celebrate your engagement, then work together for the realization of your happiness. You still have several years of hard work ahead of you. May your great love for Else give you strength for this.

Regard her now as your ideal, the striving for which will imbue you with greater energy than if you had already attained it. I hope you understand me. "Onwards!" – That must be your motto. You have been fortunate to win an outstanding girl. Now it is also your duty to make her happy, and that you can do only if, with care and deliberation, you first make provision for a nest. You are still so young.

Leontine will also be very pleased, because she loves Else as much as I do. The joyful news has increased our impatience and longing for you. If today I do not enlarge upon my happiness, please do not misunderstand me. My feeling and excitement is too great. When I am with you again and can feed on your happiness, you will see in my eyes how happy I am."

And to Else, written on the same date and likewise in German.

"My Dear Else,
If I do not at once address you as Daughter, it is because this is too ordinary for me. You are to me not what I write, but what I feel. Only when I return, then through proof of my love, I want to make the word "Mother" easier for you.

I want to be a mother to you like your mother, to care for you and strive for you. Many years of testing still lie ahead of you. I want to do all in my power to help you both attain your goal. Through steadfastness may your love for each other grow stronger. Bruno loves you. With the whole strength of his love he will work to build up your future, but too great haste undermines strength. Quietly and with determination he must strive towards his goal.

Your tenderness of heart, your nobility of souls will help him. He has a stormy temperament, but he must learn to be patient.

I hope that only a few months still separate us, and then we can enjoy your happiness to the full. Only when you feel that I am a real Mother to you, do I want you to call me "Mother." For me you are not the stranger whom my son introduced to me as my future daughter. You are the realization of my dreams and, just as one does not like to awake from a beautiful dream, so I still wish to keep the dream today until purified by circumstance, it becomes reality to me."

So worried were Helene and Heinrich about their daughter's determination to marry her impecunious suitor, who didn't even have a job that they sent her on an extended trip to Europe under the chaperonage of the solicitor, Bernard Alexander and his wife Hannah.

Figure 16. Emma with Family in Vienna. 1891
L-R: Feodor, Valerie with Leontine, Bruno seated

It is not certain how long she was away, but hardly had she returned than she and Bruno announced their engagement. Bruno had in the meantime obtained a position on one of the Witwatersrand Reef mines. Bruno and Else were married on 24th February 1907. Helga was born nine months later, almost to the day, on 27th November 1907. Else had thought they should not have a child immediately, but Bruno said:

"If two people love each other as much as we do, then the course of nature must not be interfered with."

Figure 17. Bruno when a student at Mining University in Leoben, Austria.

Figure 18. Else in Berlin shortly before getting married.

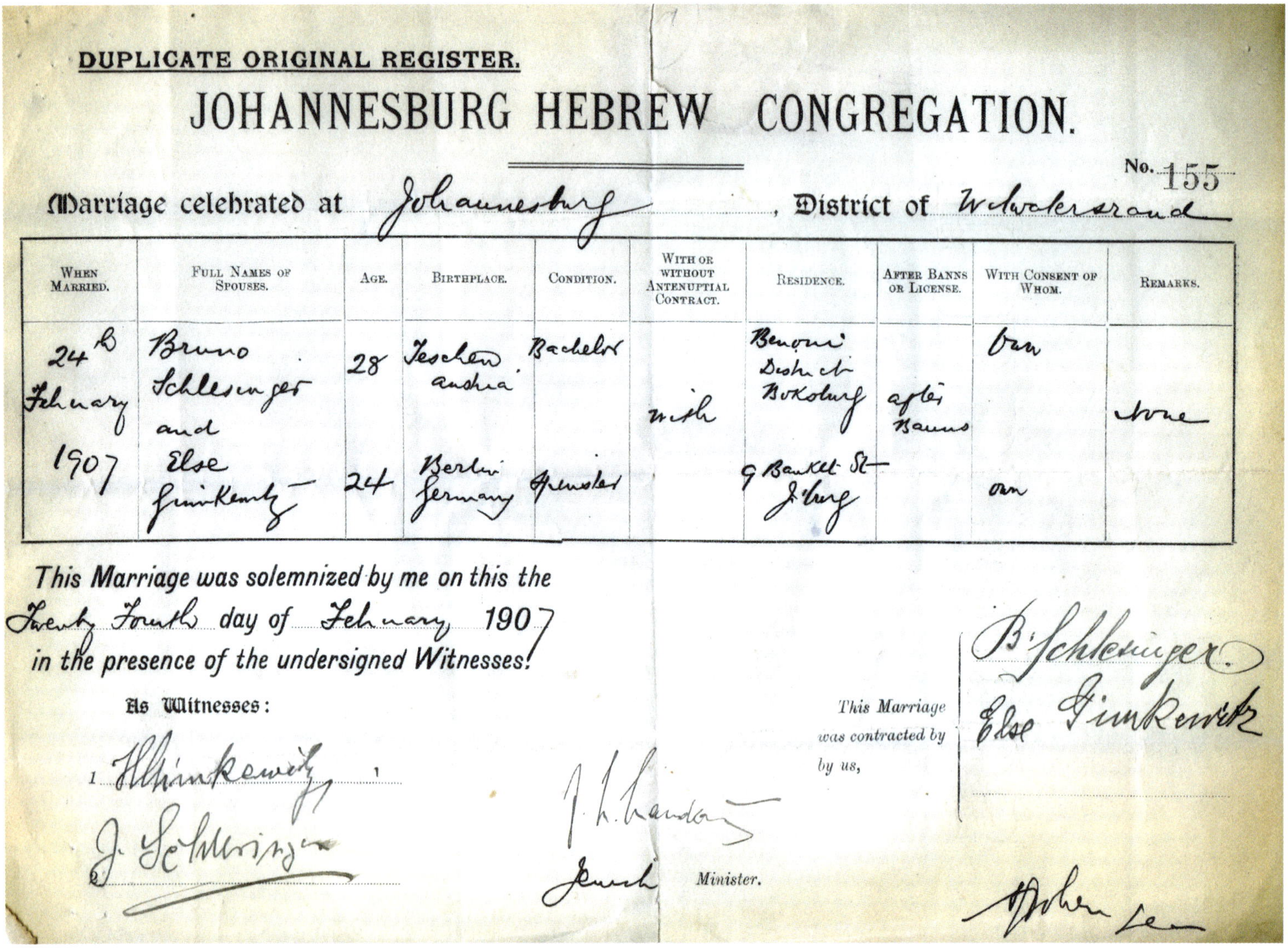

DUPLICATE ORIGINAL REGISTER.

JOHANNESBURG HEBREW CONGREGATION.

No. 155

Marriage celebrated at Johannesburg, District of Witwatersrand

When Married.	Full Names of Spouses.	Age.	Birthplace.	Condition.	With or without Antenuptial Contract.	Residence.	After Banns or License.	With Consent of Whom.	Remarks.
24th February 1907	Bruno Schlesinger and	28	Teschen Austria	Bachelor	with	Benoni District Boksburg	after Banns	Own	None
	Else Grinkewitz	24	Berlin Germany	Spinster		9 Bartlett St J'burg		Own	

This Marriage was solemnized by me on this the Twenty Fourth day of February 1907 in the presence of the undersigned Witnesses.

As Witnesses:

1. H Minkewitz
J. Schlesinger

J. L. Landau
Jewish Minister.

This Marriage was contracted by us,
B. Schlesinger
Else Ginkewitz

Figure 19. Bruno and Else Original Wedding Certificate. 1907

Figure 20. Bruno and Else Katuba. Translated as: On the First day of the week, the Tenth day of the month Adar in the year 5667 A.M. to Sunday the 24th of February 1907 the Holy Covenant of Marriage was entered into, in between the Bridegroom Bruno Schlesinger and the Bride Else Gimkewitz. The said Bridegroom made the following declaration to his Bride: "Be thou my wife according to the Law of Moses and of Israel. I faithfully promise that I will be a true husband unto thee. I will honour and cherish thee; I will protect and support thee, I will work for thee and provide all that is necessary for thy due sustenance, even as it beseemeth a Jewish Husband to do. I also take upon myself all such further obligations for thy maintenance during thy lifetime as are prescribed by our religious statutes." And the said Bride has plighted her troth unto him, in affection and with sincerity and has thus taken upon herself the fulfillment of all duties incumbent of a Jewish wife. This Covenant of Marriage was duly executed and witnessed this day according to the usage of Israel.

Map 2. South Africa with today's borders showing cities and towns lived in or visited by Schlesinger Family.

CHAPTER 3.

Early Times as Mining Engineer in God's Own Country.
1907-1910

In 1904, 50,000 Chinese labourers had been brought to Johannesburg to work in the mines, however by 1907 Blacks became accustomed to working underground and most of the Chinese were repatriated. Dad nearly fell foul of one of those who remained. Always a strict disciplinarian, he must have lost his temper with a laggard. The latter drew a knife, but before he could use it, Bruno knocked it out of his hand. None of the underground workers ever rebelled again.

When I was a few months old Dad was commissioned to prospect for gold in the untamed wilds of what was then Rhodesia (now Zimbabwe). Mom, I and my nanny Ursula, a descendant of the Incas of Peru, soon followed him. Even after the restricted life on the mine, the three thatched roof rondavels ('round houses'), far from civilization to which Bruno took his wife and child must have come as a shock to Mom. She, however, like here husband, was imbued with a spirit of adventure; any hardship and discomfort they endured only seemed to cement their love.

When my brother Rolf (born on 22nd September, 1909, like me, at the Queen Victoria Nursing Home in Johannesburg) and I grew older, Dad often regaled us with the true stories of his adventures in Rhodesia:

Once, while camping on a mining site, he was about to climb onto his stretcher, when he notice a slight movement of his pillow. Underneath it, comfortably ensconced, lay a huge python. It was soon dispatched with one shot of Dad's rifle.

Figure 21. Bruno and Else's home in Rhodesia. 1907

On another occasion he was awakened by a puffing and snorting. Leaping from his camp bed, he was just in time to see a rhinoceros stampede straight through the tent of one of his employees and carry a portion of the canvas off on his horn. The man was obviously wide-eyed and terrified, but unhurt.

Our favourite tale was of the time Dad lost his way in the bush. It was already dark and he could not risk wandering through the dense bush without a torch or firearm. The intermittent roar of a lion in the distance caught his attention and then came progressively nearer and nearer. Climbing into a tree, Bruno tied himself to a stout bough with his braces, for fear he may fall asleep, exhausted as he was after a long day. Sleep, however, was out of the question. The stillness of the night was broken by an occasional grunt as a pair of yellow eyes circled the base of the tree. Eventually the lion tired of his monotonous vigil, for when dawn broke, there he lay, peacefully asleep on the ground far below. Fortunately, he must have been well fed. A few well aimed branches accompanied by a fierce shout send him plodding off into the thicket.

When Bruno arrived at the rondavels Else flung herself into his arms. She too had spent a sleepless night, alert to every sound and fearful for her husband's safety.

Else had the ability to make a home from the meanest surroundings. The little rondavels in which they now lived were as comfortable and cozy as the houses on the various mines and as the large home, years later, in the well to do suburb of Houghton in Johannesburg, from which I was married.

Dad loved the outdoor life. Never, during his worst trials and disappointments, did he ever consider returning to Europe. South Africa was for him, "God's Own Country".

(Bruno's father Isidor died on 27 August 1910 in Johannesburg. We do not know the details. His great grandchildren, however, Erica Wolter (Wolf) and Keith Kaye did manage to find the written Johannesburg Chevra Kadisha (Jewish Burial Society) records which state "Died at 34 Leyds Str: Age 68 years." Then, by further careful searching, they found his grave in the old Braamfontein Jewish Cemetery in Johannesburg in May 2006.)

Figure 22. Else, Helga with nanny Ursula. Rhodesia. 1908

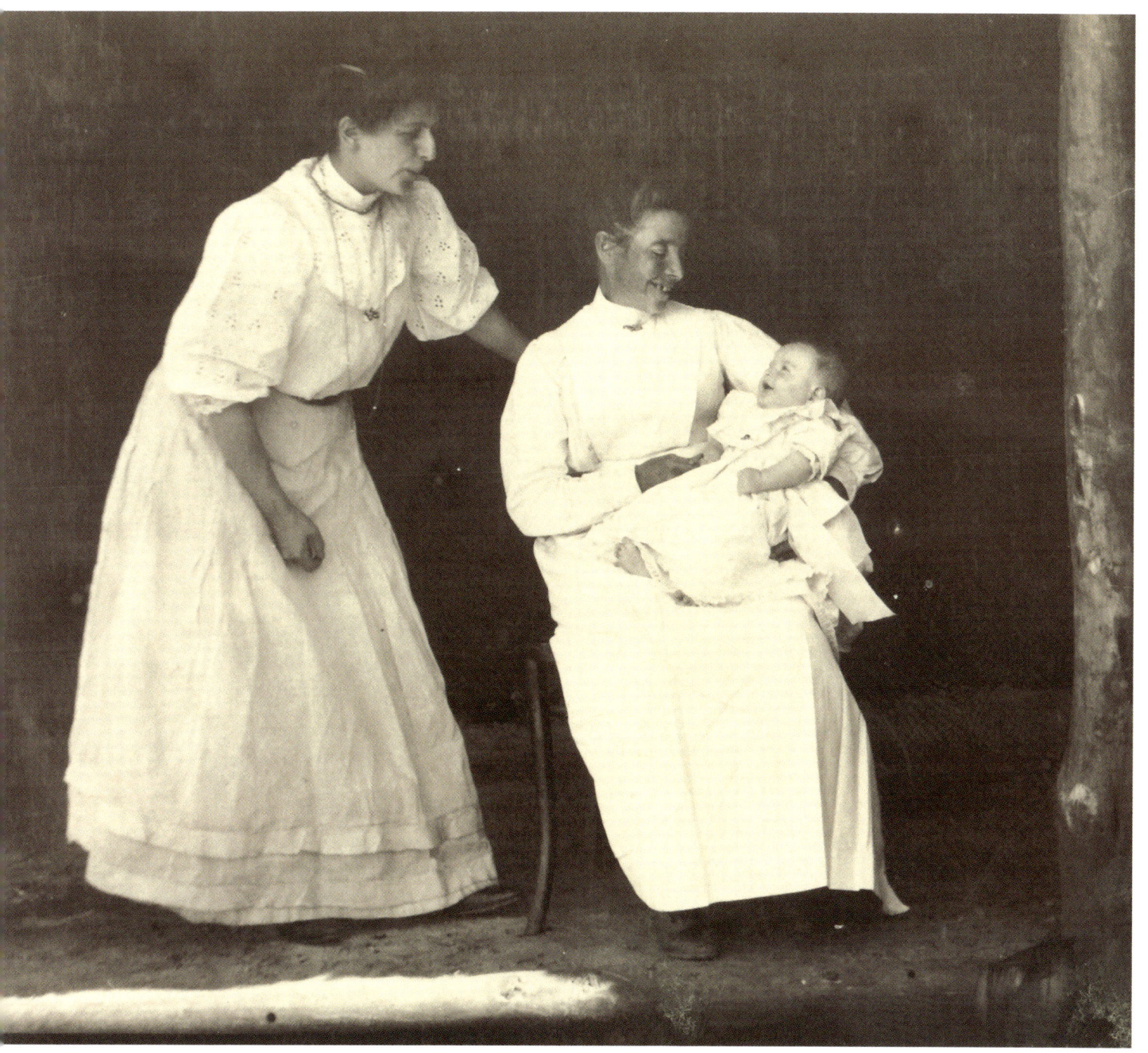

Figure 23. Else, Helga, Ursula. Rhodesia.

Figure 24. One year old Helga in Rhodesia. 1908

Figure 25. Playful Bruno.

Figure 26. Prospecting in Rhodesia.

Figure 27. Sick whilst prospecting in Rhodesia.

Figure 28. Strong man Bruno on vacation at Natal South Coast. 1910

Figure 29. Inscription from Johannesburg Chevra Kadisha (Jewish Burial Society) Records. 1910

Figure 30. Isidor Schlesinger's Grave in old Jewish Section of Braamfontein Cemetery, Johannesburg, which Erica Wolter (Wolf) and her cousin Keith Kaye had just discovered. May 2006

Figure 31. Keith and Erica at Braamfontein Jewish Cemetery. Johannesburg. May 2006

Figure 32. Schlesinger Family in their car at Main Reef Goldmine, Johannesburg. 1914. L-R: Bruno, Rolf, Else, Helga.

CHAPTER 4.

World War 1, Anti-German Sentiment, Internment, Mozambique.
1914-1917

When World War 1 broke out feelings on the mine ran high. Our parents had been very popular in the small community and had been welcomed into every social activity; however the fact that Father was Austrian assumed major proportions. This was especially among those on a lower rung of authority who were anxious to step into his shoes. Rumours, completely baseless, were eagerly fostered. We were training our Great Dane, Farkash, to bite the British, we were spies, the fan-tailed pigeons which had been given to Rolf as a birthday present, were carrying messages to the enemy. Former 'friends' (with the exception of the Hooper's) turned away when they saw our parents. One even spat in front of Mother. One evening, two plain clothes men came for Dad. As an enemy alien he was sent to an internment camp at Fort Napier in Pietermaritzburg.

Dad had suffered for a long time with his stomach and after only a few months became seriously ill. After recovering in the camp hospital he was released on parole.

On May 7th, 1915 the British ship "Lusitania" was sunk by a German submarine off the coast of Ireland. Anti-'Hun' pitch reached a frenzied pitch in South Africa as elsewhere. Anyone thought to have German sympathies was attacked in the street. The windows of the German Club were smashed and its furniture hurled out and burned. Most remaining enemy aliens were rounded up. Bruno, having had a taste of prison life, could not endure the thought of again being confined. He hid at the home of his sister and brother in law, Dr Emil Pollak. There being a dearth of doctors, members of the medical profession, who had not engaged in politics, were permitted to continue their practice. The Pollak's were friendly with a family named Frankel and when it was whispered that Jakob Frankel was going to flee to what was then Lourenco Marques (now Maputo) in Mozambique, in his chauffer driven car, Emil arranged for Bruno to accompany him.

Since having been hounded from the mine we had lived with our maternal grandparents, Heinrich and Helene Gimkewitz at 52 Esselen Street in Hillbrow, Johannesburg. Our grandfather owned the best known toy shop in Johannesburg, situated in Palace Buildings at the corner of Rissik and Pritchard Streets. He was greatly loved, not only by his many child customers and their parents, but also by the numerous friends and acquaintances that enjoyed the warm hospitality of his home. In the hysteria that swept the community after the sinking of the Lusitania however, many houses belonging to German citizens were ransacked and burned. I remember Grandpa removing the number from the gate of his house for fear that rioters might descend upon it. We children were sent to English friends for safety.

Fortunately the precautions were unnecessary. Our grandparents experienced no violence, but, being unable to import German goods and having in any case, lost the majority of his customers, the toy shop had to close its doors. From that time our grandmother was forced to take in boarders so as to maintain the home.

The separation from her husband, the uncertainty and the worry about his health preyed on Else's mind. At last, after some months, an opportunity arose for her to travel to Lourenco Marques. Leaving us children in the care of her parents she set off with a light heart. The reunion must have been a very joyous one. When she entered the single room that Bruno occupied at the small guest house however, she was puzzled to see it littered with match boxes. She picked one up.

"Don't touch that!" Dad exclaimed, "One of my best friends is confined in that box."

Mom stared at him in alarm and retreated a step. Surely the loneliness and lack of freedom for this most active, independent man, had affected his brain. For a few seconds the tension was at breaking point. Then Dad broke out into his usual boyish laughter.

"Don't worry Schatzi, I'm not mad. Come, I'll let you into my secret. Each of these boxes contains a spider…."

"A spider!" Mom shrieked.

She was not normally afraid of insects but her husband's blunt statement was confirmation of her fears for his sanity,

He patted her hand.

"Let me explain" he said. "I want to collect the fine threads they spin to use on delicate scientific instruments which now during the war can no longer be imported from Germany or even repaired"

How far he succeeded in this bizarre enterprise I do not know, but it did help to relieve his boredom.

One afternoon, several months after her return from Lourenco Marques, Else was sitting in the lounge with her parents when there was a knock on the door.

"Bruno!" Grandpa exclaimed as he opened the door.

Mom dropped her sewing and rushed to the door. There stood her husband, disheveled, unshaven, his eyes bloodshot and glazed as though in a fever.

"Quick, close the door," he whispered through cracked and swollen lips as he stepped out of the light. Else took him in her arms.

"My darling what happened? Where d'you come from? You're ill, let me help you."

"Water, I need water" he murmured.

Grandmother, who had followed the others to the door, hurried to the kitchen, while Else and her father helped the exhausted man up the stairs to bed. Before closing his eyes he clutched his wife's arm.

"Nobody must know I'm here", he said urgently.

Bruno slept for two days and two nights, waking only to drink the egg flip Mom prepared. It was Wednesday when he awoke sufficiently refreshed to tell Mom of his flight from Lourenco Marques.

Unconsciously following the route taken by nineteenth century adventurers from the coast to the goldfields in Eastern Transvaal, Dad crossed mountains, waded through rivers and swamps still infested with malaria carrying Anopheles mosquitoes and cut his way through dense forests until he reached the sleepy mining village of Pilgrim's Rest, where his father Isidor, had explored for gold so many years before. Afraid that the miners who still haunted the diggings might have been warned of his escape, Bruno camped in the forest and sent a black man, who he had befriended and who was helping to carry his few possessions, to buy fresh provisions. That was the last Dad saw of his companion or his belongings. From now on the journey was a nightmare!

On one short stretch a young couple took pity on the weary, emaciated figure plodding along the dusty road and offered him a lift in their one horse carriage, sharing their sandwiches and flask of tea with him. They were on their way to Barberton, however, which was much further south and not on the way to Johannesburg, so Bruno had to resume his journey on foot. As he neared Pretoria, he hid during the day and walked throughout the night.

All this we children were told only on the Friday afternoon, so that we would have the whole weekend to grow accustomed to the thought of having our father home. At school everyone knew that he was a prisoner of war: a careless word might send the police on his tracks.

We listened with fascination to the account of his adventures. He told how, while camping in the forests, he was sometimes bombarded with twigs by the monkeys. He still sought their company however; as he knew that at the first sign of any danger they would disperse with loud shrieks and chattering.

I adored my father. To me he was a knight in shining armour, but I was nearly instrumental in having him rearrested.

During the weekend following that on which Rolf and I had been let into the secret, I was playing in the bedroom with a school friend. There was a knock on the door.

"Come in Daddy," I called. "Only Jill is here"

My father froze. Without a word, he turned and stalked from the room. I then heard him going from room to room, and later realized this was to investigate the best means of escape should there be any attempt to re-arrest him.

"I will not, under any circumstances, allow myself to be confined again," he told my mother.

Fortunately the contingency did not arise.

CHAPTER 5.

Barman, Manager Diamond Mine and Box Factory. Return to Profession.
1917 - 1922

An influential friend in Pretoria obtained a position for him as barman in a hotel. Here he worked under an assumed name. His identity soon became known however, but through the influence of men like Sir George Albu and Sir Anthony Viljoen, who had great confidence in Dad's ability and integrity, he was in 1917 offered the managership of a small diamond mine, the New Stella, at Rayton near the Premier Mine not too far from Pretoria.

The two years we spent there were, I think, as happy for our parents, as they were for us. Rolf and I attended the little village school which boasted three class rooms for all the standards from the grades to standard five. On the two mile walk there and back I told Rolf, two years my junior, stories to shorten the distance. At one stage the government offered the use of donkeys to children living more than a mile from the school.

Rolf and I gaily climbed on our mounts, but the animals refused to budge. Our friends tried to push them, but no amount of persuasion helped. When we did eventually reach home, half pushing, half pulling, we were more exhausted than if we had walked all the way as usual. During the night one of the donkeys broke his halter and wandered away. Our parents had to pay a fine to get him out of the pound. We were all relieved when our four footed enemies were returned to the source from whence they had come.

Dad became very friendly with the Mining Commissioner, Mr. Terry, who had the finest home in the village. It was there that Rolf and I enjoyed our first games of tennis. Around this time a high smoke stack was erected on the mine. Dad was warned that this could not be permitted as the villagers believed it could be used to send wireless messages

In about January 1918, Dad became a partner in a box factory in Paarl in the Cape. This was a very unhappy time for him. He was completely out of his element. The work he did and the people he associated with were totally foreign to him. Sundays, however we all enjoyed. Before the sun shone with its normal intensity we climbed Paarl Mountain with its three majestic dolomite boulders, Paarl, Britannia and Gordon's Rocks. We eschewed the "goody-goody path as we called it and under Dad's expert guidance climbed barefoot up the steep smooth slope of Paarl Rock. At a little stream we picnicked, often stabbing crabs with a hat pin and roasting them on a fire.

As the peace bells rang in November 1918 our train puffed into Park Station in Johannesburg. Dad could no longer suffer the pettiness and intrigues of his partners in the box factory. The war was over and he could at last follow his profession of consulting geologist and mining engineer.

Miss Lienhardt owned a small house at 60 Joel Road in Berea, Johannesburg. A colleague of Bruno's, Dr Hans Merensky, had a room there. As Miss Lienhardt was going to Germany for two years Merensky suggested that we rent the house and he would continue living there. Thus began the close association, both professional and social with Hans Merensky who later became such a well known geologist.

Hans Merensky, with his modesty, his twinkling blue eyes and his good humour was an ideal boarder. He and Dad undertook many geological investigations together. Times were hard, money was scarce, but fortunately Mom was an excellent organizer and housewife. She stood by her husband in every endeavour. Her courage and strength helped him overcome many a disappointment.

The greatest of these and one that darkened his later life was his friendship with Hans Merensky. Several times when the latter was in dire financial straights, Dad lent him money which he himself could ill afford. On one occasion Bruno stood guarantor for his friend and narrowly saved himself from bankruptcy, when the enterprise on which Merensky had embarked, failed.

Figure 33. On Rayton Mine near Pretoria, with Else and Rolf. 1917

"Schlesinger," Merensky used to say when Dad was in the depths of depression, "cheer up. By the end of the year we'll make 50,000 pounds.

When some years later he discovered the Namaqualand Diamond Fields and made many times that amount, the old friendship was forgotten and Dad suddenly became a stranger to this now famous geologist. Indeed, during the Great Depression of the early 1930's, when Bruno had lost his all, Else, unknown to her husband, wrote to Merensky asking him to help tide Dad over his difficulties. Bruno's pride would never have permitted him to ask assistance of anyone for himself. Else received no reply.

This was however in the distant future. In 1922 Merensky became part of the family. He was included in all events and in every invitation to our parents. His charm and friendliness endeared him to all our relatives and friends. Every morning he, Dad and I walked to town. I was left at my school, The Commercial High School, which was on the Union Grounds, while the men continued to their offices. Rolf was still at the Yeoville Government Primary School.

Adversities and his many disappointments often caused deep depressions and skepticism in Dad's later life. As a young man however, he had all the verve, enthusiasm and zest for life which had swept Mom off her feet.

Before coming to South Africa with her parents, Else had studied languages and literature in Prague. This stood her in good stead during the years of World War 1 when Bruno was first interned and was later in Mozambique. It also helped when speculations on the Johannesburg Stock Exchange robbed him of his gains as one of the best known and most trusted mining engineers in the Transvaal. Else was able to contribute towards the cost of living by teaching French privately and, during the Great Depression of the 1930's, she taught at Kingsmead School for Girls

Dad was a strict disciplinarian. Although he was away from home so often, he was always intimately concerned with our well being and education. His great love for us did not blind him to our faults, because he was intelligent and far seeing enough to realize that without self restraint and a sense of responsibility we could never achieve lasting happiness. Good manners and a good posture were of paramount importance to him. Today I am grateful to him for his insistence that we walk and sit erect. I can still hear his angry voice at the table:

"If you don't sit up straight, I'll hammer a nail into the chair back so that it pricks you every time you slouch" and when we failed to hold our knife and fork correctly or sat with our elbows on the table he stormed:

"I could have taught a puppy how to sit and eat properly more easily than you".

He taught us not to discuss things that happened in the home with outsiders and that punctuality was an essential part of good manners. It was inconsiderate to keep anyone waiting. Once he engaged a workman for an important position on the mine. He was to be at our house at 8 am. When he hadn't arrived by 8.05 Dad drove off without him. The man lost his job.

"One cannot rely on a person who is not even on time for an appointment," he said.

He was interested in all our activities and was just as ready to join us at play, as to help and advise us with our homework. His greatest relaxation was in the open air. He disliked inactivity and his greatest joy was to walk in the country. To go 10-12 miles on foot to a picnic spot and back was a panacea for him.

We stood in awe of our father and would never have dared to argue with him, as we did with our mother. Nevertheless, we did love and admire him. Now, from this distance however, I realize he was often too strict with Rolf. He believed a man must be tough and thus in his dealings with him, may have robbed Rolf of some self confidence. He always treated me, as a female, courteously and at times was decidedly over indulgent.

In some respects he was unbelievably naïve. He did not like Mom or me to use cosmetics other than a light powder. Once, when traveling by car, we passed a woman at a bus stop. She obviously was heavily rouged and wore bright red lipstick. Turning to me Dad said:

"You see what a lovely complexion that girl has? She doesn't spoil it with all kinds of powders and paints"

Nor was I allowed to wear earrings, because it looked "common". Once Josse and I were engaged however, and my fiancé gave me a pair of marquisite baubles Dad never again voiced any objection.

```
7154 - 14/10/20 - 2000        Ex Unitate Vires.        2/6 Revenue Stamp
                                                            Cancelled.
 C                                                            1/8/21
   O
     P         UNION  OF  SOUTH  AFRICA.
       Y.
```

C E R T I F I C A T E   O F   N A T U R A L I Z A T I O N.

No. 7 2 5 8.

WHEREAS Bruno Schlesinger a native of Austria, and an Austrian subject by birth and at present residing at Johannesburg in the Province of the Transvaal, Union of South Africa, has applied for a Certificate of Naturalization in the terms of the Naturalization of Aliens Act, 1910;

AND WHEREAS the said applicant has complied with the provisions of the above-named Statute, and intends, when naturalized, to continue to reside within the Union of South Africa.

AND WHEREAS notice of the intention of the said applicant to apply for a Certificate of Naturalization has been published in the Gazette;

AND WHEREAS the said applicant has made and subscribed the declaration of allegiance to His Majesty the King;

NOW THEREFORE, under and by virtue of the powers conferred upon me by the said Act, I do hereby grant this Certificate of Naturalization to the said

BRUNO SCHLESINGER

and I do further declare that he shall, except as is otherwise provided by law, be henceforth entitled to all the rights, powers, and privileges, and be subject to all obligations to which a natural-born British subject is entitled or subject in the Union of South Africa.

Given under my hand at PRETORIA this fifteenth day of August, 1921.

(Signed) H. VENN,

Under Secretary for the Interior,
for Minister of the Interior.

Figure 34. Copy First Naturalization Certificate. 1921

Figure 35. Doing what he loved best – out in the open veld, prospecting and exploring. 1921

Figure 36. Prospecting with Else. 1921

CHAPTER 6.

Life as Mining Engineer.
1924 - 1934

Bruno was a man of the strictest integrity. A Mr. P Smit asked him to take charge of his diamond claims during his absence overseas, as he was "the only honest man" he knew. Dad, however, was no business man. Mom, had a better insight into human nature. Time and again she warned him against ventures with men in whom she had no confidence, but he believed everyone to be as honest and straightforward as he was. When Mom's intuitive premonitions proved to be correct, he sank into the deepest depression. This, however, did not last long. He was, above all a man of action. When one venture failed, he was soon ready to devote his never flagging energy to another.

For years he was associated with the financier, Jimmy Kapnek and shared his confidence that one day oil would be found at Inyaminga in Mozambique. In Potgietersrust and Lydenburg he worked with Merensky developing the platinum fields. In 1924 there was a platinum boom and it was thought that Lydenburg would become a second Johannesburg. A year later, however, the boom somersaulted to a slump.

December 1924 was the start of a momentous time for diamonds in South Africa and Bruno was soon to be involved in the middle of the turmoil. John Voorendyk, the postmaster of Lichtenburg, a small village about 230 Km West of Johannesburg, in what is now the North West Province of South Africa, was digging a hole for a cattle dip. His cattle were dying of some unknown disease. As the hole was nearing completion one of the workers caught sight of a shiny object in the ground which he showed to John's son Koosie. "Dad! Look! A Diamond." Not sure exactly what they had found they took the stone to Mr. Bosman, the local science teacher. He put the stone through a test of acid and confirmed that this was in fact a real 3 carat diamond.

For some time there was doubt whether this was a significant find. In 1925 other diamonds were found on some of the surrounding farms and then in February 1926 the farm Elandsputte was the first to be proclaimed a Diamond Mine. This started off a flurry of activity with thousands of professional diggers, fortune seekers and adventurers flocking to the area. They came from everywhere, around South Africa, Europe, Australia and arrived by car, bicycle, on foot, in donkey carts and ox wagons. All with one word in their minds. Diamonds!

Dad now became involved both in the prospecting of the region and running of the mines. On the farm Welverdiend, where more than 1.5 million carats of diamonds were to be mined, he erected the mine headgear for the owners, Messrs Donaldson and Carlis, and he supervised the mining as long as it was still considered profitable. In 1928 this led to a very complicated law suit by Bruno against the owners. (See Appendix 8)

Dad spent several years working in the Lichtenburg area, coming to Johannesburg only occasionally and here in 1927 he confirmed the presence of diamonds on the farm Grasfontein. Here the biggest 'Diamond Rush' in world history took place on the 4th March 1927. During this time we spent our school holidays with him and I was fortunate enough to witness this greatest, and one of the last "Diamond Rushes" in the world. Would be diggers either ran for themselves, or engaged professional runners, to stake their claim. At a prearranged signal they set off from the starting line in a flurry of excitement to stake their 31 foot by 31 foot claim.

Altogether more than 2 million carats were mined from Grasfontein. From the Lichtenburg area between 1926 and 1945, 7 million carats were found with a value in today's terms of R2000 Million (US$ 300 Million). These massive diamond finds led Sir Ernest Oppenheimer to establish the Oppenheimer Syndicate which bought up huge quantities of the diamonds and then let them trickle slowly back onto the world markets, thus preventing a total collapse of the world diamond market.

Figure 37. My Dad. Enjoying the outdoors. 1924

On the 15th July 1932 (a month after Josse and I were married and living in Klerksdorp) he wrote to us from "White Limes" in Potgietersrust: "It's just like old times…. I had rather a gigantic task here, but since today the back of the job is broken and tonight I shall be able to go to bed at 9.30 instead of, as heretofore, at midnight….But there is nothing like a difficult job well carried out. It makes life worthwhile, no matter how small the salary, no matter how severe the work."

And on 22nd November 1932: "Things here in Potgietersrust are running so smoothly, both in the mine as well as the factory that I am fed up to the hilt. There's nothing to put my fists into (my teeth having departed long ago). All hard nuts have been cracked and the organization which I myself have sired, is now maddening me with its silky purr. From morning to eve I dash about looking for trouble, and not a blasted one wants any. And yet I can't get away, for there are always mice about."

Figure 38. Last page of 1928 Welverdiend lawsuit, Schlesinger v Donaldson and Carlis with Bruno's signature.

Our mother bore his frequent, often lengthy absences from home with fortitude and outward calmness, although she missed him more than she would admit. She loved him with all the intensity of her warm, sentimental nature and worried about him when he was unwell or far away in the wilds. She would have been happy to accompany on his many expeditions, where his profession of necessity took him, however she conceived it as her duty to make a home for us children to whom she was a selfless loving friend and companion as well as a mentor. She was not only with us, but of us, a confidante with whom we could share our most intimate thoughts and problems.

She bore with poise and equanimity her husband's occasional bursts of temper. Bruno was quick to arouse, especially when beset by worries, but his ill humour did not last long. He never sulked or bore a grudge.

Shortly after Josse and I were married, he wrote to us: "I am unspeakably happy at your happiness, you Two! If ever you have a tiff (the salt of married life), never go to bed without making up. Never!! I am telling you, who knows."

A great factor in their marriage was their consideration for each other, their ready appreciation of even the smallest pleasures and, above all, their enthusiasm.

I must quote from a letter of Mom's written to us in Klerksdorp: "….. and now my great surprise. Yesterday I was about to leave for town when the bell rings. Lifting his hat and with a deep bow, stands by the hall window --- our Daddy, What do you think I did? Not go to town, oh no! until that afternoon when we both went together." Their realization of the true values of life and their sense of humour helped them over many a difficulty.

Owing to his contacts in Mozambique and the high opinion the authorities had of him, Dad was invited to Portugal on two occasions. In 1930, and again in 1937 when he was commissioned by Salazar, President of Portugal to inspect and report on silver mines north of Lisbon that had been worked on by the ancient Romans. From Portugal he visited us in London, where Josse was studying medicine and we spent happy weeks together. All our friends were enchanted with his charm, vivacity as well as his prowess on the dance floor.

Figure 39. Last great "Diamond Rush" Grasfontein. March 1927

Figure 40. Enlargement Grasfontein Diamond Rush and boy with rabbit.

Figure 41. Mines developing at Grasfontein.

Figure 42. Diamond washing equipment at Grasfontein.

Figure 43. Else and Helga at primitive diamond sorting table. Grasfontein. 1927

SCHLESINGER, Bruno, Mining Engineer, Consulting Mining Engineer and Geologist; b. 22nd March, 1879, Troppan, Silesia. Educ. Vienna and Austria; m. Came to South Africa 1900. Mem. Scientific and Technical Club, Johannesburg. Rec.: Sport. Add., 4, Sherbourne Road, Parktown, Johannesburg.

Figure 44. Entry from Who's Who in South Africa. 1927

Figure 45. Portrait around 1928.

Figure 46. Bruno Schlesinger Family about 1928.
L-R: Else, Rolf, Helga, Bruno.

Figure 47. With wife, mother and son. About 1929.

Figure 48. Santo da Serra, Portugal. 7th July 1930

Figure 49. Sao Vicente, Portugal. 1930

Figure 50. Wedding Helga and Josse. 49 St. Patrick's Road, Houghton, Johannesburg. 19 June 1932. L-R: Bruno, Fritz Gimkewitz, Sophie Gimkewitz, Josse and Helga Kaye, Lily Pollak, Else Schlesinger, Rebecca and Meyer Lichtenstein.

Figure 51. Helga, Rea, Bruno. Klerksdorp. 1934

Figure 52. Bruno, with his niece Dr. Ilse Schlesinger (Wolf). Italy. 1934

No. 14263

UNION OF SOUTH AFRICA.

British Nationality in the Union and Naturalization and Status of Aliens Act, 1926.

CERTIFICATE OF NATURALIZATION

Granted to a Person who was Naturalized before the passing of the above-mentioned Act.

Whereas .... BRUNO SCHLESINGER ....

of .... Johannesburg ....

in the Province of .... the Transvaal .... Union of South Africa, being an alien who was naturalized in the Union of South Africa or in a Province now included in the Union of South Africa before the passing of the above-mentioned Act, has applied for a Certificate of Naturalization, alleging with respect to .... himself .... the particulars set out below, and intends to continue to reside within the Union of South Africa:

And whereas the Minister is satisfied that such a certificate may properly be granted:

And whereas the said .... BRUNO SCHLESINGER ....
has made and subscribed the Oath of Allegiance to His Majesty the King, His Heirs and Successors:

Now, therefore, in pursuance of the powers conferred on him by the said Act, the Minister grants to the said .... BRUNO SCHLESINGER .... this Certificate of Naturalization, and declares that the said BRUNO SCHLESINGER shall, subject to the provisions of the said Act, be henceforth entitled to all political and other rights, powers, and privileges and be subject to all obligations, duties, and liabilities to which a natural-born British subject is entitled or subject; and have to all intents and purposes the status of a natural-born British subject.

By order of the Minister.

DEPARTMENT OF THE INTERIOR,
CAPE TOWN: 9th March, 1934.

Secretary for the Interior.

Signature of holder .... B Schlesinger

Signed in my presence ....
(Insert Official Designation.)

ELECTORAL OFFICER
JOHANNESBURG

Place ....
Date ....

Particulars Relating to Holder.

Full name	BRUNO SCHLESINGER
Address	49, St. Patrick Road, Houghton, Johannesburg
Trade or occupation	Mining Engineer
Place and date of birth	Troppau, Czecho-Slovakia; 22nd. March, 1879
Nationality	(original) Austrian
Married, single, or widower (widow)	Married
Name of wife	Else Gimkewitz
Names of parents	Isidor Schlesinger and Emma Fasal
Nationality of parents	Austrian
Date and place of previous naturalization	15th August, 1921; Pretoria
Act under which previously naturalized	Act No. 4 of 1910

Figure 53. Naturalized as British Subject in South Africa. 1934

CHAPTER 7.

Family. A Good Friend.
1935 - 1942

In his profession Dad was greatly respected, not only for his knowledge and skill, but above all, for his integrity and reliability. Had he invested his earnings in guilt edged securities, he would have saved himself much anxiety and become a wealthy man, but unfortunately he played the Stock Exchange. When the crash came from 1929 to 1932, he lost his all. The large fully furnished house at 49 St. Patrick's Road, in Houghton, Johannesburg which he had bought as a surprise for Mother when she and I returned from two years in Europe at the end of 1929 (he had always hoped to join us, but was prevented by last minute business commitments) had to be sacrificed to pay his debts. After that our parents always lived in rented homes

For Josse and me he was the best friend we ever had, always ready to help even at great personal sacrifice. As soon as he heard that Josse had taken ill in 1942, Dad came to Cape Town to help us, although he had only recently recovered from a heart attack himself.

He arranged with his friend Friedrich Knacke that we should occupy his mansion "Earls Dyke" in Camps Bay free of charge. The surroundings were beautiful with balconies north, south, east and west. Here it was hoped Josse could catch the sun at all times of the day and this would help him recover from the tuberculosis to which he had fallen prey after working 18 hours a day for the past year.

Josse had no income and we had no assets other than our few sticks of furniture. With two young children our future looked bleak. Dad was in a very precarious financial position himself. His ill health and the slump of the Stock exchange had eaten into his earnings. At great cost to his pride, and only because it was not for himself, he induced Josse's and my family to contribute towards a monthly income for us until such time as Josse recovered. The doctors were adamant that Josse could not resume general practice. He tried again to get into the army but was turned down because of his health. When a course in Radiology was instituted at Groote Schuur Hospital in Cape Town, Josse registered for this specialty. Thanks to Dad's efforts our families continued to help us until such time as Josse was well enough to resume his career.

Dad was a good friend, not only to us, but to all who needed him. When World War 11 broke out Knacke was to be interned. Dad went to see General J.C. Smuts and pointed out how frail and sickly his friend was. He said he would guarantee that Knacke would not take part in any political activities. (Dad himself had been naturalized after World War 1). Smuts agreed not to have Knacke interned, but said he must leave "Earl's Dyke which overlooked the sea, as he was concerned that signals may be sent to enemy ships or submarines. Knacke was to live inland for the duration of the hostilities. He went to Worcester where he had to report to the police daily until, through Dad's further efforts this was reduced to first weekly and then monthly visits to the police.

At "Earl's Dyke" Dad felt responsible for his friend for the care of the fifteen room mansion. We were inundated with visitors and weekend guests, and, although we had some of Knacke's servants to help us, it was not easy to cope. Dad devoted himself to the garden, cutting fresh steps, sawing timber and building a retaining wall at one of the swimming pool. His energy was boundless despite the precarious condition of his heart.

We had spent barely six months at "Earl's Dyke", when we received a message that Crown Prince (later King) Constantine and Princess Frederike of Greece were coming to inspect the house with a view to renting it. Dad at once contacted Knacke to undertake to pay the costs of transport and our hotel expenses until we were able to find alternate accommodation.

After weeks of procrastination the Greek Royal Couple decided against renting "Earl's Dyke", because most of their entourage lived in the suburbs. By that time we had, however, already found a flat in Arlington Court, Muizenburg.

Dad and Mom now returned to Johannesburg where Dad found that he had missed several valuable commissions through his prolonged absence. (He never mentioned this us; his diaries revealed the true state of affairs).

(Bruno's Mom, Emma died on the 18th August 1940, at the Florence Nightingale Nursing Home in Johannesburg. Interestingly his wife Else was born on the same day in 1882, and his daughter Helga died on that day in 1998. Somewhat mysteriously, despite knowledge of her death and having her death certificate, the family has not been able to locate where she is buried.)

Figure 54. With friends.

Figure 55. At home in Houghton, Johannesburg, with Emma, and Else. 2 November 1935

Figure 56. Picnic with Granddaughter Rea-June Kaye. Epping Forest, London. February 1939

Figure 57. Great Friends.

Figure 58. London May 1939. Portrait by son-in-law, Dr Josse Kaye.

2869/40

DEATH NOTICE

PURSUANT TO THE PROVISIONS CONTAINED IN "THE ADMINISTRATION OF ESTATES ACT, 1913."

1. Name of the deceased .... Emma Schlesinger
2. Birthplace and Nationality of the deceased .... Wirombani Austria Cechoslovak
3. Names and Addresses of the Parents of the deceased — Father: Josef Fasal — Mother: Charlotte b. Newmann
4. Age of the deceased .... 86 years .... 1 .... months.
5. Occupation in life of the deceased, or if a woman, of her husband ....
6. Ordinary place of residence of the deceased, or, if a woman, of her husband .... 34 Washington Court Fortes Ave. Yeoville, Johannesburg
7. Married or unmarried, widower or widow ....
   - (a) Name of surviving spouse (if any), and whether married in community of property or not ....
   - (b) Name or Names and approximate date of death of pre-deceased spouse or spouses .... Isidore Schlesinger 27 Aug 1910
   - (c) Place of last marriage .... Teschen Austria
8. The day of the decease: On .... 18 August 1940
9. Where the person died — House: Florence Nightingale Nursing Home — Town or place: Johannesburg — District: ....
10. Names of children of deceased, and whether majors or minors .... Valerie Pollak; Feodor Schlesinger; Bruno "; Leontine Fleisher-Sagan

State separately the children born of different marriages, and give the date of birth of each minor. Names must be written out in full. If there are no children, and either or both parents be dead, then give the names and addresses of the brothers and sisters of the deceased.

11. Has the deceased left any movable property? .... ✓
12. Has the deceased left any immovable property? .... ✗
13. Is it estimated that the estate exceeds £300 in value? .... ✗
14. Has the deceased left a will? .... Y

Dated at Johannesburg .... the 26 day of August 1940

(Signature) B Schlesinger
Son

(State in what capacity and whether at the time at or near the place of death).

This notice must be filled in and signed by the Nearest Relative or Connection of the Deceased, who shall, at the time, be at or near the place of death — or in the absence of such near Relative or Connection, by the person who at or immediately after the death, shall have the chief charge of the house in, or the place on which the death occurred, and must be sent either to the Master, or, if the death occurred in a district wherein a seat of provincial government is not situate, the Magistrate of the District, in duplicate within fourteen days of the death.

AURORA/10-38/11984

Estate worth £38

Figure 59. Emma Schlesinger's Death Certificate.

Figure 60. With Grandson Colin Schlesinger. 1942

Figure 61. With Grandson Keith Woodhill Kaye at 'Earls Dyke" near Cape Town.
February 1943

CHAPTER 8.

Illness and Heart Attack.
1943-1945

Early in April 1943, Dad complained of a continuous pain in the right side of his chest. It was diagnosed as being due to a 'nerve', probably as the result of sitting in a draught. A fortnight later a specialist discovered that he had a 'spasmodic heart'. From 15th June to 7th August he was in bed with asthma. He was very depressed and wrote in his diary:

"Damage to the heart. I suppose what it comes to in practice is that my life is over. That being so, the quicker the better. I don't want my poor Else tortured."

He could not however, bear to be fussed over. He, so proud and independent wrote:

"I feel the chains to my freedom. Every move of mine is carefully plotted and watched….

I am always on the defensive in a suppressed rage…."

In September 1943, our parents spent a month in Durban in the hope that change of altitude and the sea air would help Dad's heart and asthma. In spite of his weakness and ill health he still busied himself with his sister Medi's and our financial affairs. He felt no better physically or spiritually for the holiday. He was a man of action and chafed at the enforced idleness:

"Unless I find work soon, or work finds me, I shall go dotty."

On 22nd March 1944, his 65th birthday he wrote in his diary:

My birthday – probably the last one. No loss." And later:

"I should like to see the children in Muizenberg and say in my thoughts goodbye to them."

Mom and Dad came to Muizenburg shortly afterwards. In spite of frequent ill health Dad drove into Cape Town with Josse, who had started the course in radiology. There he had meetings with the many financiers and mining men with whom he had associated. Rolf attended to his business interests in Johannesburg, as far as possible.

When he felt up to it he still did physical exercises every morning and took long walks along the beach. He was also an avid and selective reader and liked to discuss the books he had read.

In October 1944 he was commissioned by a Mr. C. Verster to inspect an asbestos property at Elandsheuvel near Barberton. The prospect of work gave him a short, new lease on life. He traveled from Cape Town to Johannesburg and then on to Barberton in the Eastern Transvaal. From there he rode on horseback into the mountains and worked in the intense heat for three and a half hours without feeling undue fatigue. Unfortunately he found the occurrence of asbestos to be very restricted and reported that the property needed opening up. (His report, in his beautiful copper plate handwriting is in the Geological Museum in Johannesburg.

Shortly after his return to Muizenburg he was again ill, but forced himself t take a daily walk on the promenade. On 11th January 1945 he was invited to play bridge at the home of his friend, Meyer Mendelssohn in St. James. Mom stayed home with us. At about 10 pm there was a phone call;

"Your father has taken ill and we're sending him home with our chauffeur"

Mom, Josse and I hurried down the three flights of stairs (there was no lift) to await Dad's arrival. Before helping him up the many steps Josse gave him the glass of brandy he had brought down with him. Once upstairs Josse gave him an injection. It was of no avail. Dad's heart had finally given in.

Mom bore the shock with her usual self control. She had to be sedated for several nights, but always anxious to avoid upsetting us, especially as Josse had only returned from hospital after another operation a few days previously. At night, when she was in bed however, I often heard her stifled sobs. She lived with us for the next seventeen years until her death on 12th July 1962.

Figure 62. Last photograph. Muizenberg. 1944

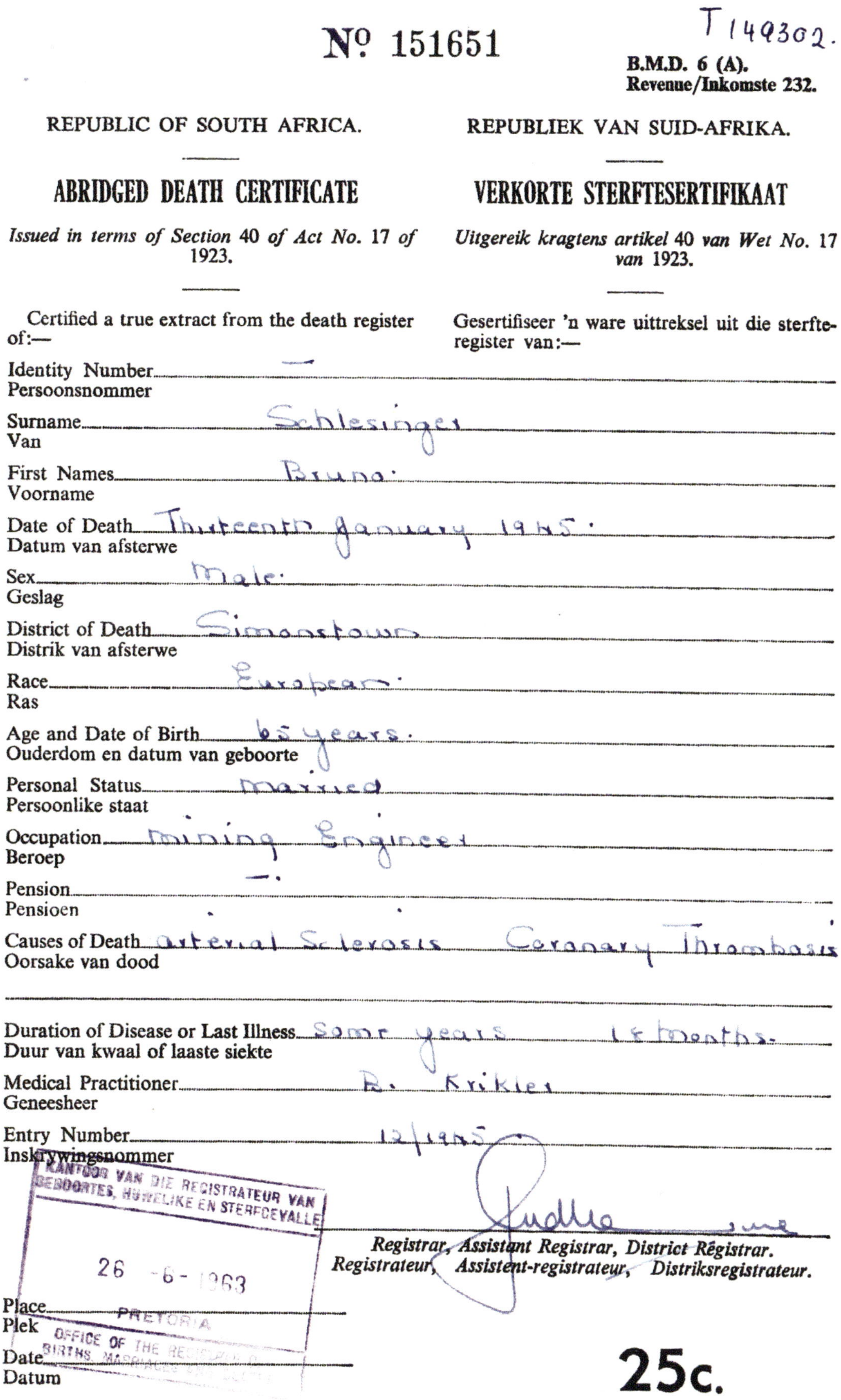

T 149302.

№ 151651

B.M.D. 6 (A).
Revenue/Inkomste 232.

REPUBLIC OF SOUTH AFRICA. | REPUBLIEK VAN SUID-AFRIKA.

ABRIDGED DEATH CERTIFICATE | **VERKORTE STERFTESERTIFIKAAT**

Issued in terms of Section 40 *of Act No.* 17 *of* 1923. | *Uitgereik kragtens artikel* 40 *van Wet No.* 17 *van* 1923.

Certified a true extract from the death register of:— | Gesertifiseer 'n ware uittreksel uit die sterfte-register van:—

Identity Number / Persoonsnommer: —
Surname / Van: Schlesinger
First Names / Voorname: Bruno.
Date of Death / Datum van afsterwe: Thirteenth January 1945.
Sex / Geslag: Male.
District of Death / Distrik van afsterwe: Simonstown
Race / Ras: European.
Age and Date of Birth / Ouderdom en datum van geboorte: 65 years.
Personal Status / Persoonlike staat: married
Occupation / Beroep: mining Engineer
Pension / Pensioen: —.
Causes of Death / Oorsake van dood: arterial Sclerosis Coronary Thrombosis
Duration of Disease or Last Illness / Duur van kwaal of laaste siekte: Some years 18 months.
Medical Practitioner / Geneesheer: B. Krikler
Entry Number / Inskrywingsnommer: 12/1945.

KANTOOR VAN DIE REGISTRATEUR VAN GEBOORTES, HUWELIKE EN STERFGEVALLE
26 -6- 1963
PRETORIA
OFFICE OF THE REGISTRAR OF BIRTHS, MARRIAGES AND DEATHS

Registrar, Assistant Registrar, District Registrar.
Registrateur, Assistent-registrateur, Distriksregistrateur.

Place / Plek
Date / Datum

25c.

Figure 63. Abridged copy of Death Certificate.

Figure 64. Rea Gardy at her Grandparents grave. Braamfontein Cemetery. Johannesburg. 14 November 1999

Figure 65. Bruno, Else Tombstone.

APPENDIX 1.

Descendant Tree of Isidor Schlesinger

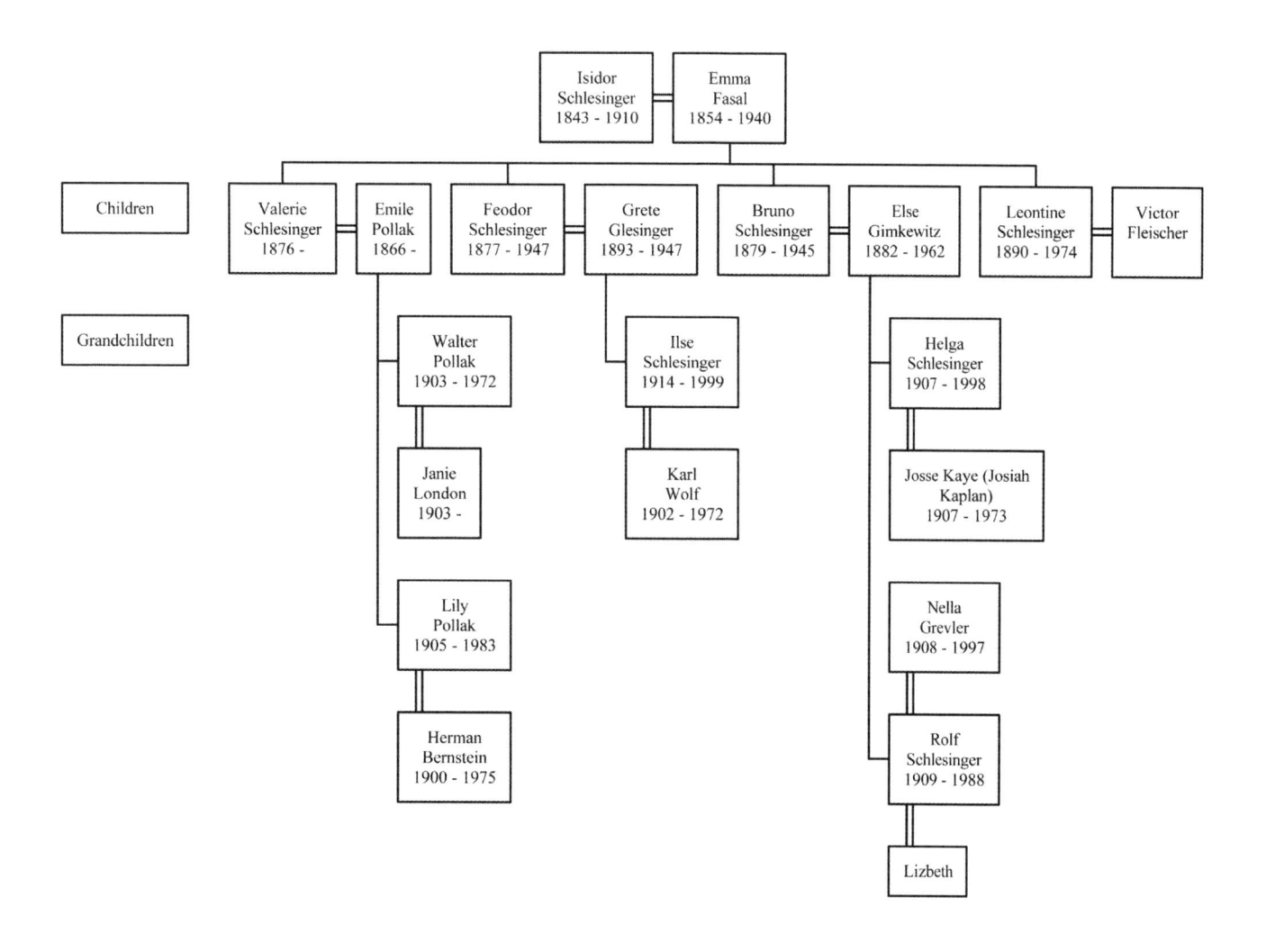

Isidor Schlesinger 1843 - 1910 — **Emma Fasal** 1854 - 1940

Children:
- Valerie Schlesinger 1876 - — Emile Pollak 1866 -
- Feodor Schlesinger 1877 - 1947
- Grete Glesinger 1893 - 1947
- Bruno Schlesinger 1879 - 1945
- Else Gimkewitz 1882 - 1962
- Leontine Schlesinger 1890 - 1974 — Victor Fleischer

Grandchildren:
- Walter Pollak 1903 - 1972 — Janie London 1903 -
- Lily Pollak 1905 - 1983 — Herman Bernstein 1900 - 1975
- Ilse Schlesinger 1914 - 1999 — Karl Wolf 1902 - 1972
- Helga Schlesinger 1907 - 1998 — Josse Kaye (Josiah Kaplan) 1907 - 1973
- Nella Grevler 1908 - 1997 — Rolf Schlesinger 1909 - 1988
  - Lizbeth

APPENDIX 2.

Hourglass Tree of Bruno Schlesinger

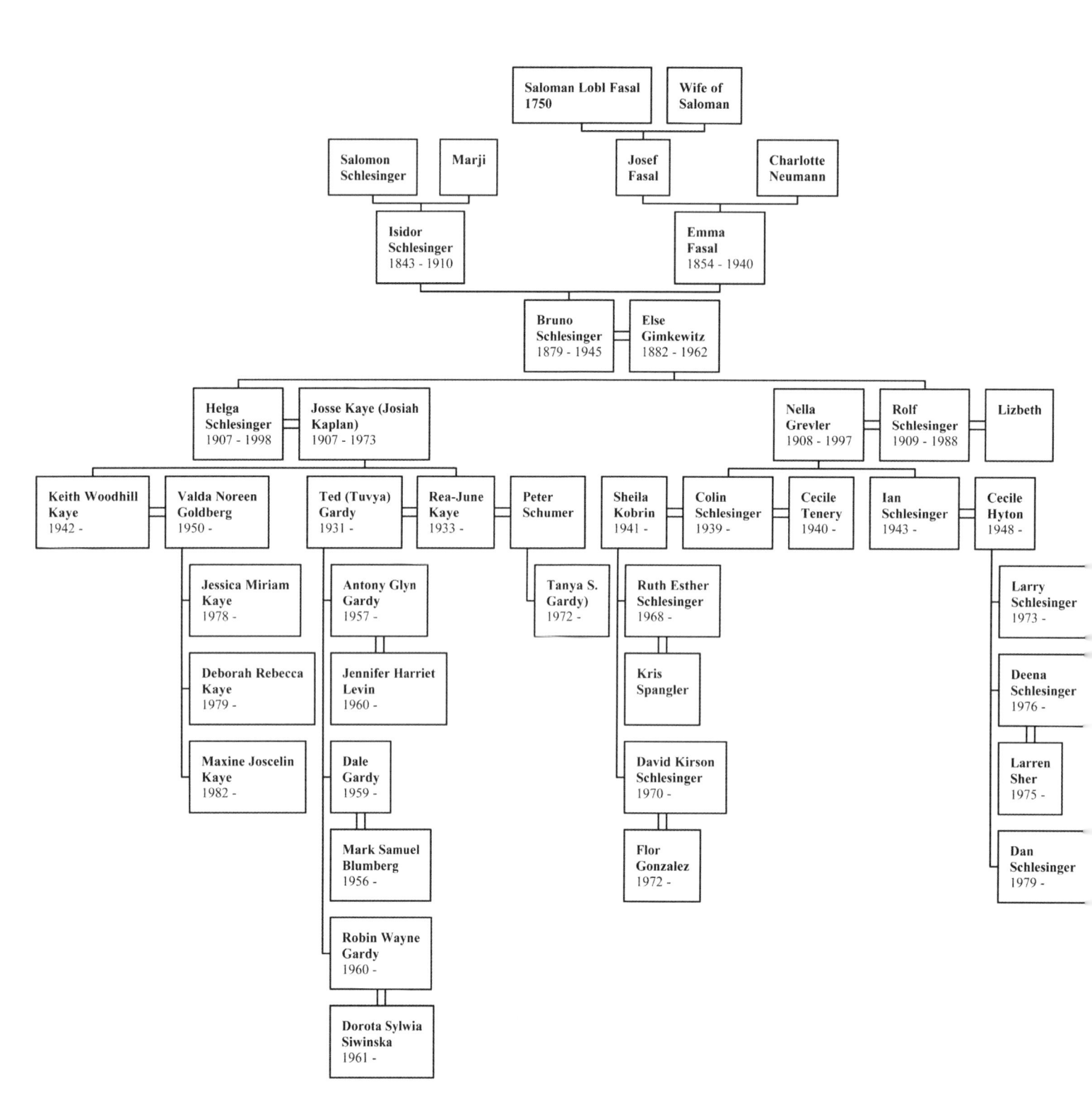

```
                              ┌─────────────────┬─────────────┐
                              │ Saloman Lobl Fasal │  Wife of    │
                              │      1750          │  Saloman    │
                              └─────────────────┴─────────────┘
                                          │
    ┌─────────────┬─────────────┐    ┌─────────────┬─────────────┐
    │  Salomon    │   Marji     │    │   Josef     │  Charlotte  │
    │ Schlesinger │             │    │   Fasal     │   Neumann   │
    └─────────────┴─────────────┘    └─────────────┴─────────────┘
              │                                │
         ┌─────────────┐                ┌─────────────┐
         │   Isidor    │                │    Emma     │
         │ Schlesinger │                │    Fasal    │
         │ 1843 - 1910 │                │ 1854 - 1940 │
         └─────────────┘                └─────────────┘
                         │
              ┌─────────────┬─────────────┐
              │    Bruno    │     Else    │
              │ Schlesinger │  Gimkewitz  │
              │ 1879 - 1945 │ 1882 - 1962 │
              └─────────────┴─────────────┘
```

Children of Bruno Schlesinger & Else Gimkewitz:

- **Helga Schlesinger** 1907 - 1998 — **Josse Kaye (Josiah Kaplan)** 1907 - 1973
- **Nella Grevler** 1908 - 1997 — **Rolf Schlesinger** 1909 - 1988 — **Lizbeth**

Next generation:

- **Keith Woodhill Kaye** 1942 -
- **Valda Noreen Goldberg** 1950 -
  - Jessica Miriam Kaye 1978 -
  - Deborah Rebecca Kaye 1979 -
  - Maxine Joscelin Kaye 1982 -
- **Ted (Tuvya) Gardy** 1931 - — **Rea-June Kaye** 1933 -
  - Antony Glyn Gardy 1957 - = Jennifer Harriet Levin 1960 -
  - Dale Gardy 1959 - = Mark Samuel Blumberg 1956 -
  - Robin Wayne Gardy 1960 - = Dorota Sylwia Siwinska 1961 -
- **Peter Schumer**
  - Tanya S. Gardy) 1972 -
- **Sheila Kobrin** 1941 -
  - Ruth Esther Schlesinger 1968 - = Kris Spangler
  - David Kirson Schlesinger 1970 - = Flor Gonzalez 1972 -
- **Colin Schlesinger** 1939 -
- **Cecile Tenery** 1940 -
- **Ian Schlesinger** 1943 -
- **Cecile Hyton** 1948 -
  - Larry Schlesinger 1973 -
  - Deena Schlesinger 1976 - = Larren Sher 1975 -
  - Dan Schlesinger 1979 -

68

APPENDIX 3.

Descendant Tree of Helga Schlesinger (Kaye)

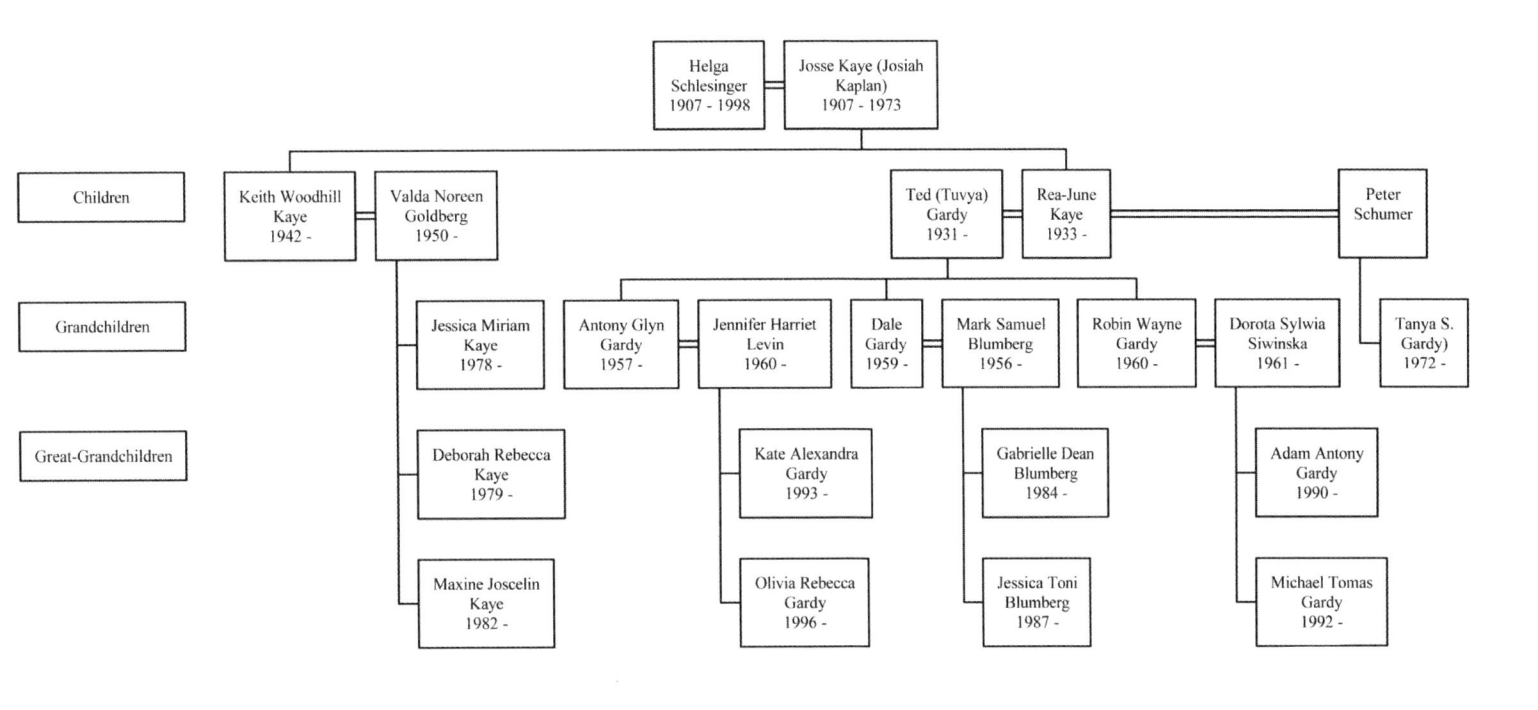

```
                          Helga          Josse Kaye (Josiah
                       Schlesinger            Kaplan)
                        1907 - 1998          1907 - 1973
```

Children:
- Keith Woodhill Kaye (1942 -) = Valda Noreen Goldberg (1950 -)
- Ted (Tuvya) Gardy (1931 -) = Rea-June Kaye (1933 -) = Peter Schumer

Grandchildren:
- Jessica Miriam Kaye (1978 -)
- Antony Glyn Gardy (1957 -)
- Jennifer Harriet Levin (1960 -)
- Dale Gardy (1959 -)
- Mark Samuel Blumberg (1956 -)
- Robin Wayne Gardy (1960 -)
- Dorota Sylwia Siwinska (1961 -)
- Tanya S. Gardy) (1972 -)

Great-Grandchildren:
- Deborah Rebecca Kaye (1979 -)
- Maxine Joscelin Kaye (1982 -)
- Kate Alexandra Gardy (1993 -)
- Olivia Rebecca Gardy (1996 -)
- Gabrielle Dean Blumberg (1984 -)
- Jessica Toni Blumberg (1987 -)
- Adam Antony Gardy (1990 -)
- Michael Tomas Gardy (1992 -)

APPENDIX 4.

Descendant Tree of Rolf Schlesinger

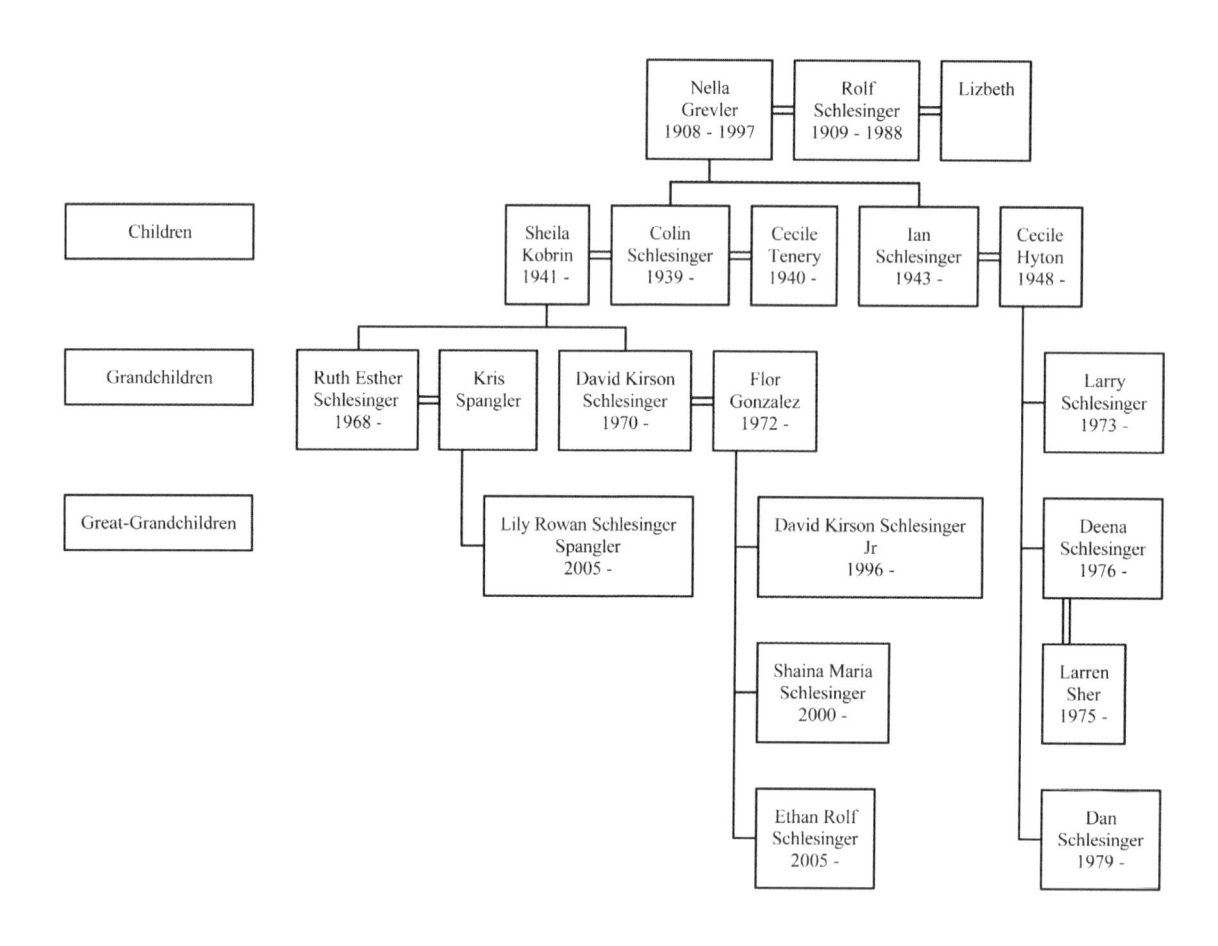

Children

- Nella Grevler (1908 - 1997) — Rolf Schlesinger (1909 - 1988) — Lizbeth
  - Sheila Kobrin (1941 -) — Colin Schlesinger (1939 -)
  - Cecile Tenery (1940 -)
  - Ian Schlesinger (1943 -)
  - Cecile Hyton (1948 -)

Grandchildren

- Ruth Esther Schlesinger (1968 -) — Kris Spangler
- David Kirson Schlesinger (1970 -) — Flor Gonzalez (1972 -)
- Larry Schlesinger (1973 -)
- Deena Schlesinger (1976 -) — Larren Sher (1975 -)
- Dan Schlesinger (1979 -)

Great-Grandchildren

- Lily Rowan Schlesinger Spangler (2005 -)
- David Kirson Schlesinger Jr (1996 -)
- Shaina Maria Schlesinger (2000 -)
- Ethan Rolf Schlesinger (2005 -)

APPENDIX 5.

Descendant Tree of Feodor Schlesinger

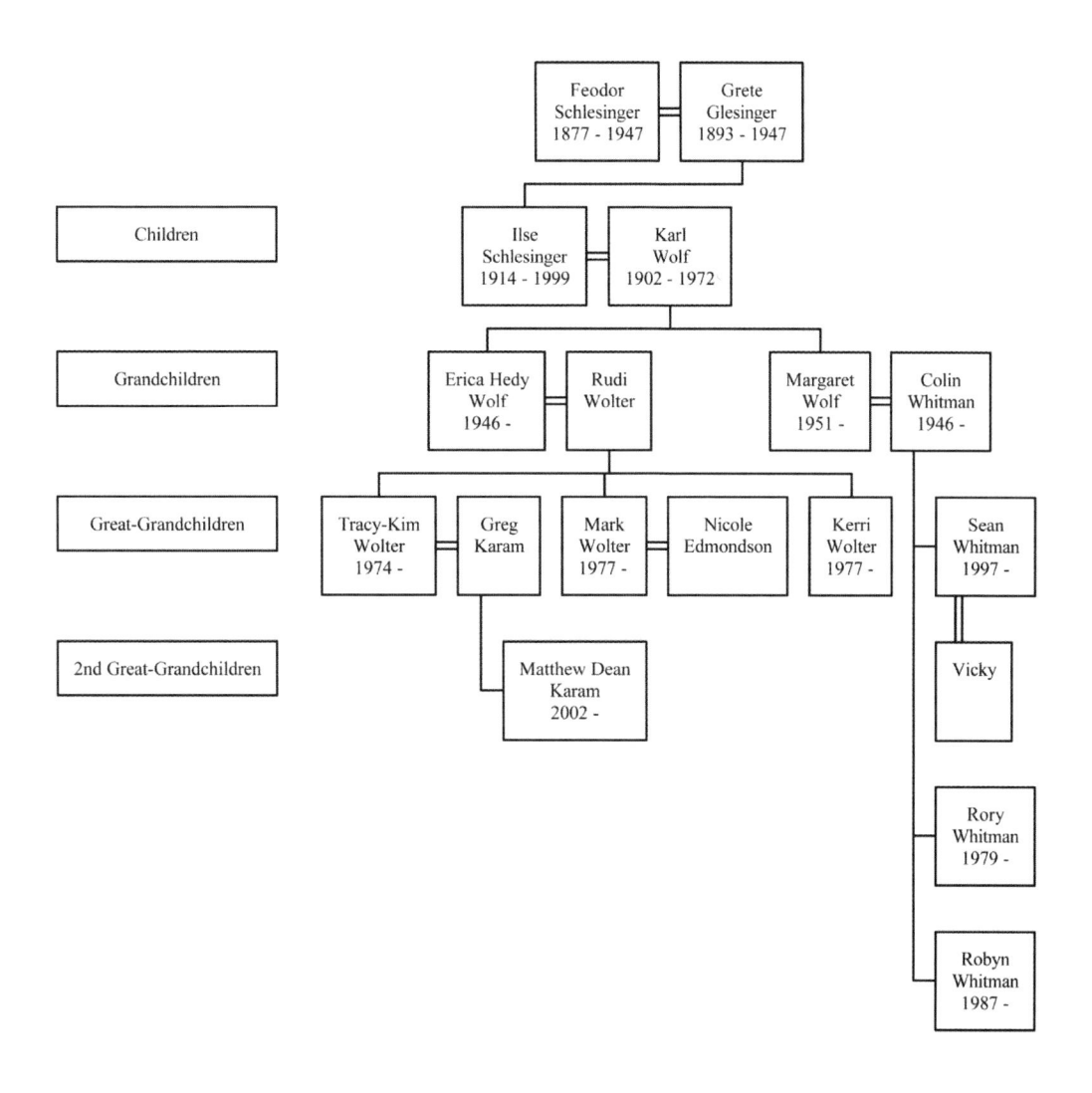

Children

- Feodor Schlesinger 1877 - 1947
- Grete Glesinger 1893 - 1947
- Ilse Schlesinger 1914 - 1999
- Karl Wolf 1902 - 1972

Grandchildren

- Erica Hedy Wolf 1946 -
- Rudi Wolter
- Margaret Wolf 1951 -
- Colin Whitman 1946 -

Great-Grandchildren

- Tracy-Kim Wolter 1974 -
- Greg Karam
- Mark Wolter 1977 -
- Nicole Edmondson
- Kerri Wolter 1977 -
- Sean Whitman 1997 -
- Vicky
- Rory Whitman 1979 -
- Robyn Whitman 1987 -

2nd Great-Grandchildren

- Matthew Dean Karam 2002 -

APPENDIX 6.

Descendant Tree of Valerie Schlesinger (Pollak)

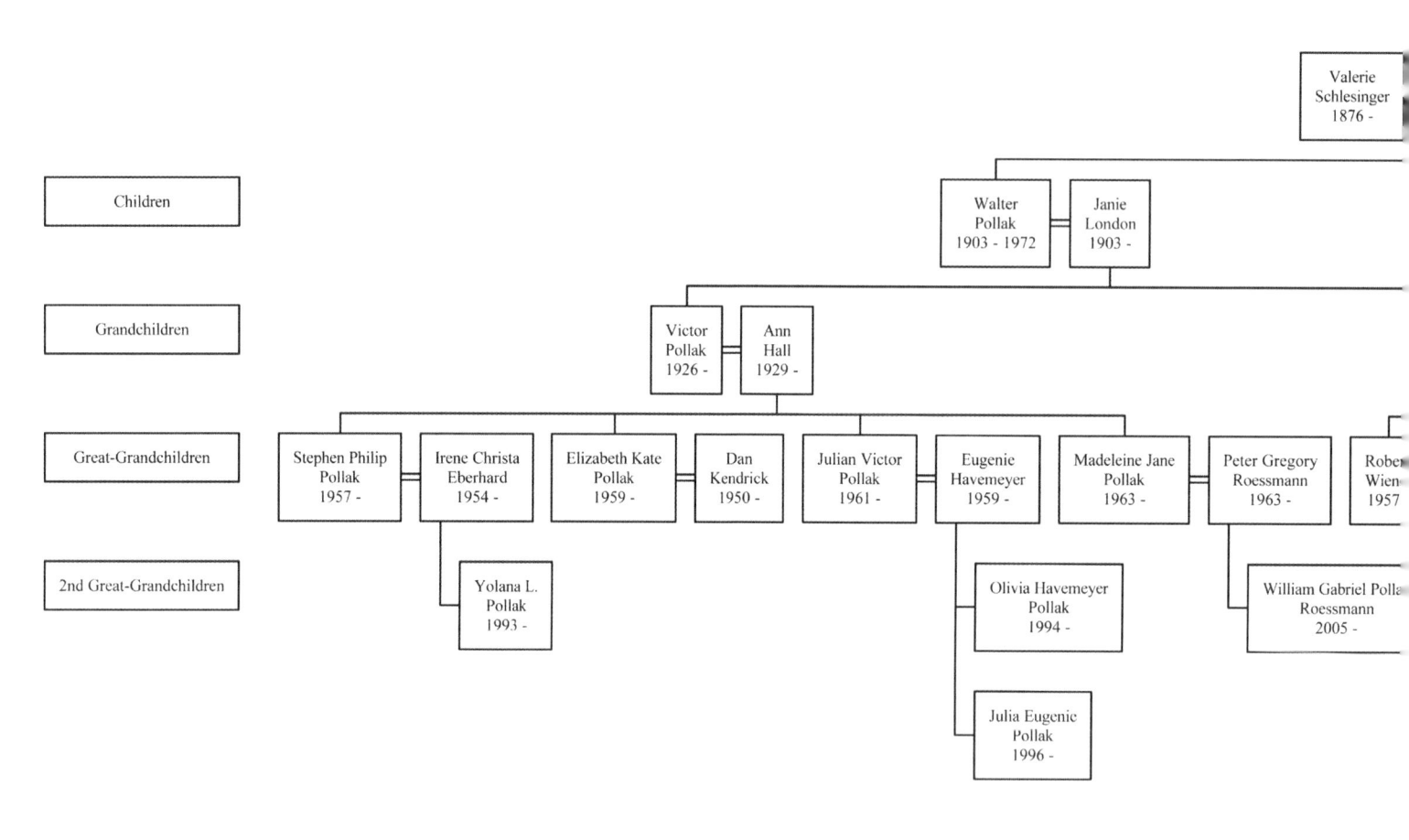

	Valerie Schlesinger 1876 -
Children	Walter Pollak 1903 - 1972 — Janie London 1903 -
Grandchildren	Victor Pollak 1926 - — Ann Hall 1929 -
Great-Grandchildren	Stephen Philip Pollak 1957 - ; Irene Christa Eberhard 1954 - ; Elizabeth Kate Pollak 1959 - ; Dan Kendrick 1950 - ; Julian Victor Pollak 1961 - ; Eugenie Havemeyer 1959 - ; Madeleine Jane Pollak 1963 - ; Peter Gregory Roessmann 1963 - ; Robert Wien 1957 -
2nd Great-Grandchildren	Yolana L. Pollak 1993 - ; Olivia Havemeyer Pollak 1994 - ; Julia Eugenie Pollak 1996 - ; William Gabriel Pollak Roessmann 2005 -

Family Tree

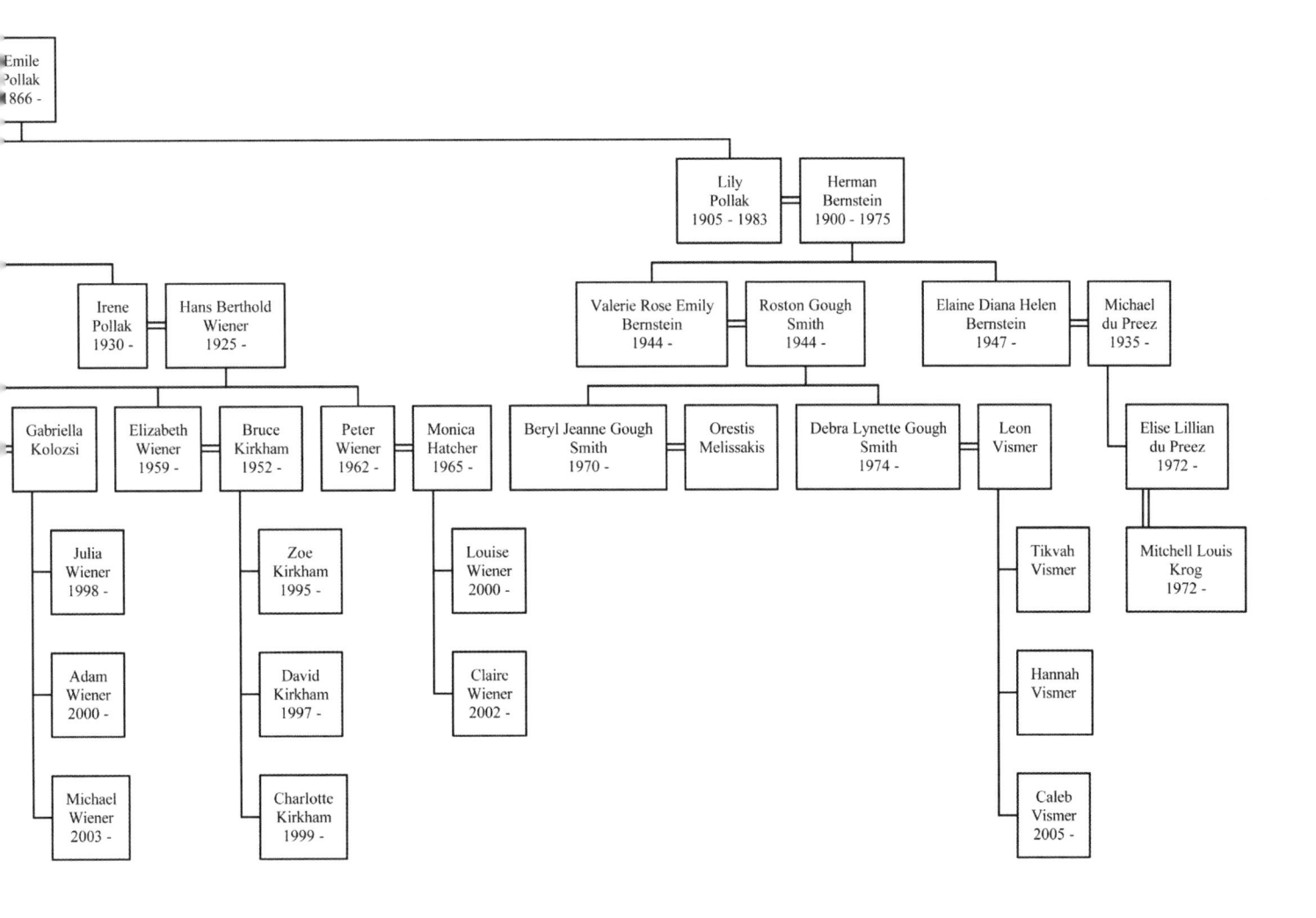

- **Emile Pollak** 1866 -
  - **Lily Pollak** 1905 - 1983 ═ **Herman Bernstein** 1900 - 1975
    - **Valerie Rose Emily Bernstein** 1944 - ═ **Roston Gough Smith** 1944 -
      - **Beryl Jeanne Gough Smith** 1970 - ═ **Orestis Melissakis**
      - **Debra Lynette Gough Smith** 1974 - ═ **Leon Vismer**
        - **Tikvah Vismer**
        - **Hannah Vismer**
        - **Caleb Vismer** 2005 -
    - **Elaine Diana Helen Bernstein** 1947 - ═ **Michael du Preez** 1935 -
      - **Elise Lillian du Preez** 1972 - ═ **Mitchell Louis Krog** 1972 -
  - **Irene Pollak** 1930 - ═ **Hans Berthold Wiener** 1925 -
    - ═ **Gabriella Kolozsi**
      - **Julia Wiener** 1998 -
      - **Adam Wiener** 2000 -
      - **Michael Wiener** 2003 -
    - **Elizabeth Wiener** 1959 -
    - **Bruce Kirkham** 1952 -
      - **Zoe Kirkham** 1995 -
      - **David Kirkham** 1997 -
      - **Charlotte Kirkham** 1999 -
    - **Peter Wiener** 1962 - ═ **Monica Hatcher** 1965 -
      - **Louise Wiener** 2000 -
      - **Claire Wiener** 2002 -

73

APPENDIX 7.

Law Suit Isidore Schlessinger: 1866

Affidavit from Barnett Elsner, in case Lewis and Lichtenstein v Isidore Schlessinger. The latter sued a Richard Rutherford in his own name rather than that of Elsner. Presumably in relation to a case in which Rutherford owed monies to Lewis and Lichtenstein.

saith, that, on or about the ninth day of April 1866, this deponent sent the said note to Capetown to Mr. Isidore Schlessinger, and requested him to take the necessary steps to protest the same, and recover the amount thereof, and that deponent was informed thereafter that proceedings were taken against the said Rutherford, by the said Schlessinger.

And this deponent further saith, that the said bill is deponent's own *bona fide* property, and that no one has an interest in the same but himself, and that the said Isidore Schlessinger only acted as his agent in the matter; and, having no power of attorney from deponent, at the time it became due, and the case being urgent, he summoned the said Rutherford, in his own name.

Sworn at Beaufort, on the 29th day of June 1866
Before me
J Kinneer
J.P.

Barnets Ebner

APPENDIX 8.

Law Suit: Schlesinger v Donaldson and Carlis 1928

IN THE SUPREME COURT OF SOUTH AFRICA WITWATERSRAND LOCAL DIVISION JOHANNESBURG.

In re:

    BRUNO SCHLESINGER

                                  Applicant.

    versus

    JAMES DONALDSON and WOLF *CARLIS*

                                  Respondents.

In the matter of

    BRUNO SCHLESINGER

                                  Plaintiff.

    versus

    JAMES DONALDSON and WOLF CARLIS

                                  Defendants.

TO THE HONOURABLE THE JUDGE PRESIDENT AND OTHER THE HONOURABLE JUDGES OF THE SUPREME COURT OF SOUTH ARICA.

THE Petition of BRUNO SCHLESINGER humbly sheweth:

1.   Your Petitioner is the Plaintiff in the above action.

2.   ON the 24th day of September, 1928, this Honourable Court at the instance of Plaintiff granted an order directing Defendants to make discovery under oath of the documents relating to any matters in question in the above action in the possession or power of Defendants.

3.   IN purported pursuance of such Order, Defendant DONALDSON purporting to act on behalf of himself and WOLF CARLIS, an the 5th November 1928 made discovery by Affidavit of certain documents alleged to be in his possession or power, declaring that he has not now and never had in his possession, custody or power or in the possession, custody or power of his Attorney and Agent or any other person an his behalf any documents or copy of or extract from any documents, relating to the matter in question other than and except the documents set forth in the first and second schedules attached to the Affidavit. Copy of the Discovery Affidavit is attached marked "A"•

4. THAT the Affidavit fails to make discovery of any documents in the possession or power of the Defendant WOLF *CARLIS or of his* attorney or agent or other person within the meaning of the Order of this Honourable Court.

5. THAT the Affidavit is not exhaustive of the documents that are or that have been in the possession, custody or power of the Defendants or one or other of them, or that are or have been in the possession, custody or power of their attorney and agent or other person on their behalf, and your Petitioner submits that from the facts hereinafter related there is a reasonable probability, presumption or suspicion that the defendants have in their custody or power or in the custody or power of their attorney or agent or other person on their behalf, documents other than those disclosed, and relevant to the matters in question in the action, within the meaning of Rule 52, and that your Petitioner is entitled to further and better discovery.

6. THAT your Petitioner craves leave to refer to the pleadings on the subject of the matters in issue between the parties.

7. THAT your Petitioner by letters from *his* attorney, copies Whereof are attached marked "B" and "C", has *called upon the* Defendants to make further and better discovery, but the Defendants *have* failed to comply with such demand, and your Petitioner attaches hereto, marked "D" and "E°" respectively, copy of letter from defendants' attorneys with the reply of your Petitioner's attorney.

8. THE facts referred to in paragraph 5 hereof are the following:(a) Two of the portions into which the farm WELVERDIEND was divided in freehold were named "P" and "E" portions. These portions, together with 11 other portions were sold by the defendants to the Welverdiend Diamonds Ltd., on the 5th January 1927 for £5250. 0. 0. The Deed of Sale has been disclosed by the defendants.

On the 3rd January 1927 a Discoverer's Certificate was issued to the Defendant CARLIS for 50 Discoverer's Claims on P portion and on the 14th February 1927 for 50 Discoverer's Claims on E portion. From a letter disclosed by defendants dated 11th January 1927 addressed to the African & European &: Investment Co. Ltd., there was a recovery of 8800 1/2 carats of diamonds recovered from P portion during December 1926 having an approximate value of £17,000. .On application by the African & European & Investment Co. Ltd., for their percentage of the proceeds of the diamonds recovered, to which they were entitled pursuant to their contract of sale of the farm Welverdiend to the defendants in the first instance, the defendants on the 13th January 1927 replied that "the diamonds found on Welverdiend in December were found on blocks after their sale, and no percentage of the profits accrued to us." Save as to the provisions in Clause 6 of the Deed of Sale by defendants to the Welverdiend Diamonds Ltd., reading: "That all income of whatsoever nature (except income from Trading Rights) as from 15th December 1926 shall accrue and belong to the purchaser," no discovery has been made of any contract of sale of any discoverer's or owner's rights or claims to the Welverdiend Diamonds Ltd. Portions P and 1+: were transferred to the Welverdiend Diamonds Ltd., on the let February ' 1927. On the 14th April 1927 an owner's Certificate was issued to Donaldson & Carlis (the defendants) for 200 Owner's claims on P portion, and on the 8th June 1927 a like certificate was issued to the Welverdiend Diamonds Ltd., for 200 Owner's claims on L portion.

On the 9th November 1927 application was made to the Mining Commissioner at Klerksdorp by J. A. Campbell (then Manager for Welverdiend Diamonds Ltd.,) and Marks, attorney for defendants, for transfer of the following claims to the following persons, namely:

200 Owner's claims on Portion E, from the Welverdiend Diamonds Ltd., to Isaac Bonner.

200 Owner's claims on Portion 2 of Portion V from the R.U.V. Syndicate (Proprietary) Ltd., to James Angus Campbell.

200 Owner's claims on portion U from the R.U.V. Syndicate (Proprietary) Ltd., to Wolf Carlis.

200 Owner's claims on Portion P from Wolf Carlis and James Donaldson to James Donaldson.

200 Owner's claims on Portion S. from Wolf Carlis and James Donaldson to John Murray.

These applications were accompanied in each case by Declarations of Seller and Purchaser, which as the transfers were refused were returned to applicants. Discovery has not been made of these Declarations and if the applications were in writing, of the application.

On or about the 11th May 1928 in a debate in Parliament the Honourable the Minister of Mines stated:

"Now, in regard to the Carlis Syndicate, no such body exists. What happened was this. In view of the specific provision of Section 73 of the pet, that no natural person may hold claims in trust for a corporate body, we, being under the impression and we had good reason to think so, that Carlis was merely a trustee for a corporate body, notified him that certain discoverers' claims standing in his name were virtually the property of a corporate body, and-that therefore, he should stop working them at once. He stopped working. We similarly gave notice to Carlis and Donaldson that certain owners' claims registered in their name were the property apparently and evidently of a partnership, which is also prevented from holding claims under Section 73 of the Act. We said: "You must stop the work." They stopped. Afterwards I got letters from their lawyers. Carlis said. "I am a natural person, and "do not hold these claims in trust; from the very start they have been in my name; I am willing to lay before you my books and documents and you can send anybody to investigate. Donaldson took up the same attitude. They said they were not a partnership, but held these claims as joint holders. As a matter of fact, Carlis, in regard to the discoverers' claims had anticipated the difficulties which could be created, in respect of the corporate body in which he was interested, by the new Act, and on November 7, before the new Act came into force, he entered openly into a deed of sale with that corporate body. Discoverers' claims have always been in the name of a natural person. lie laid before me the advice of various prominent counsel in Johannesburg I went into the matter and consulted the Government Attorney, and there were persons who investigated the actual facts for us, and I came to the conclusion that it was doubtful that the Government could make out a good case in a court of law, and recognised his title, and the same with Donaldson. Donaldson and Carlis resumed under the arrangement with the Government, which is the same as with the Merensky syndicate in Namaqualand."

The letters from defendants' lawyers, defendants' books and documents, the Deed of Sale, statements of 'case and Counsel's opinion and the arrangement of defendants with the Government referred to in the Minister's statement have not been disclosed.

At the time mining operations were carried on on P portion, one CAMPBELL was the manager of the Company at Welverdiend. He had authority to operate upon the banking account of the Welverdiend Diamonds Ltd. The proceeds of diamonds sold were paid into the banking account of the Company. From time to time after making provision for working expenses and other outgoings such proceeds were paid by the Company's cheque to the defendant DONALDSON personally. These payments must be reflected in the banking account of the defendants or one or other of them. Disclosure has not been made of such banking account.

After the coming into operation of Act 44 of 1927 defendants carried on mining operations on P portion and extracted diamonds. Disclosure has not been made of the number of diamonds extracted and the value thereof.

From a profit and loss account of the Welverdiend Diamonds Ltd., dated 31st December 1927, the Company appears to have realised from diamond sales, commissions, claims and other items a sum of 281,861 Pounds. The balance sheet discloses a profit of £46,050.

(b)   Certificates for 50 Discoverer's claims on each of the following portions were granted to the defendant CARLIS between the 11th January 1927 and the 14th February 1927, namely: the remaining extent of Welverdiend, portions K,U, portion 2 of portion V, portion 1 of portion A, portion 2 of portion A, portion 3 of portion A and a certificate for 200 Owner's claims was on the 25th April 1927 issued to the defendants in respect of portion K.

(c)   On the 24th November 1926 portion A and the remaining extent were sold to the Pretoria East Diamonds Ltd., for £2000. Save as to a provision contained in paragraph 7 of the Deed of Sale to the effect: "We (that is the defendants) remain in technical possession of the ground until transfer but will allow diamonds to be recovered by you under our licence and in our name before transfer on the condition that the value less cost of winning is devoted to the payment for the ground," there is no reference to any sale or disposal of any discoverer's claims.

(d)   On the 5th January 1927 the defendants sold to one S. J. Chandler portion K for 750 Pounds. the sale to take effect as and from the 17th December 1926. Save as to a statement contained in paragraph 5 that all income of whatsoever nature except income from Trading Rights as from the 15th December 1926, shall accrue and belong to the purchaser, there is no record of any sale of any prospector's or owner's claims. From discovery made by the defendants 396 carats of diamonds were recovered from S portion during December. On the same date as the Deed of Sale between Chandler and the defendants was executed, the 5th January 1927, Chandler re-sold Y. portion to the Welverdiend Diamonds Ltd., for 12,750 Pounds.

(e)   On the 2nd February 1927 defendants sold to the R.U.V. Syndicate (Proprietary) Ltd., Portions R, U and V. No record of any sale of any discoverer's claims appears in the Deed.

(f)   On the 10th March 1927 the Pretoria East Diamonds Ltd., sold to the Welverdiend West Alluvials Ltd., portions 1, 2 3, 4 and 5 of portion A and the remaining extent of the farm for 26250 Pounds., purchased on the 24th November 1926 for 2000 Pounds.

(g)   On or about the 15th May 1928 the Lichtenburg Gravels Ltd., purchased the total shareholding of the R.U.V. Syndicate (Proprietary) Ltd., for £60,000.

9.   A Company known as NEW MINES LIMITED was incorporated on the 25th April, 1925. It was then a private company and was converted into a public company in about May 1927. The Company in question is the parent company for all the defendants' ventures and transactions and practically the whole of the issued share capital of such Company is held by the defendants or members of their families or other persons on their behalf, and in as far as the transactions of the defendants in relation to the farm Welverdiend or the proceeds of any diamonds recovered or assets sold or dealings in shares may have taken place, such transactions will be reflected either in the books of the defendants or in the books of the New Mines Ltd.

10.   On the 30th December 1926 the Company Welverdiend Diamonds Ltd., was formed. It was a public company with a capital of 25,000 Pounds divided into 100,000 shares of 5/- each. The first allotment was for 76,000 shares of which 24,000 were allotted to Carlis. Of the other allotments your Petitioner has reason to believe that, if not all, the large majority of such allotments were to nominees of the defendants. A further allotment of 24,000 was subsequently made whereunder 22,800 were allotted to the defendant Carlis and the remaining allotments to persons whom your Petitioner has reason to believe were again nominees of the defendants. Although on the balance sheet lodged there has been a profit of £46,050. for the year ending 31st December 1927, there has been no declaration of dividends.

11.   ON the 24th January 1927 the Company Lichtenburg District Diamond Gravels (Proprietary) Ltd., was formed. It was subsequently converted into a public company. At the time of its formation two shares were subscribed for by two clerks in the employ of the defendants' attorneys, each of which clerks

took one share. On an allotment made for the period 25th January 1927 to 19th February 1927 of 272,000 shares, 7992 were allotted to the defendant Carlis, 24,000 to the defendant Donaldson, a further 24,000 to the Defendant Carlis, 52,000 to your Petitioner on behalf of the defendants, and 12,000 to the auditor of the defendants, and 48,000 to one Sievewright whom your Petitioner has reason to believe is a nominee of the defendants.

12. On the 1st February 1927 the R.U.V. Syndicate (Proprietary) Ltd., was formed, again subscribed for as to one share each by two clerks in the employ of the defendants' attorneys. On the pleadings it is admitted that such Company is controlled by the defendants.

13. On the 2nd March 1927 the Welverdiend West Alluvials Ltd., was registered as a public company with a capital of £50,000., divided into 200,000 shares of 5/- each. On the return of allotments made 200,000 shares were allotted between the 3rd March 1927 and the 7th April 1927. Of these shares 60,000 shares were allotted for a consideration other than cash. But for 100 shares issued to each of the subscribers to the Memorandum, 99,700 shares were issued to one Edward Morley and 99,600 shares to one James Butler, clerks in the employ of the defendants or New Mines Ltd.

14. The defendants have had substantial dealings in the shares in all the above named Companies, all of which companies were concerned with some portion or other of the farm Welverdiend, and the defendants have already and still hold substantial interests in such Companies, either in their own names or in the names of persons on their behalf.

15. Under the Agreement of the defendants with the African & European Investment Company Ltd., for the purchase of the farm, the sellers under such Deed were, in addition to the purchase price, entitled to one-half of all nett profits that may accrue from any sale, lease or other disposal of or dealing with the farm or any portion thereof, or the owner's share of mining licences, or the sale, lease or other disposal of or dealing with trading and business rights, grazing rights, water, owner's and discoverer's rights, from the carrying on of mining or prospecting operations, and from any sale, lease, disposal or other dealing with precious stones, and generally a half share of any net profit derived by the defendants from the said farm.

16. YOUR Petitioner has reason to believe that the purchaser, Chandler, of portion K was a nominee of the defendants.

17. THAT from the defendants intimate relationship with the companies that may have been carrying on mining operations on any portions of the farm Welverdiend, returns will or may have been sent to the defendants as to precious stones recovered and as to the returns made to the Diamond or other Government Department.

18. THAT in the process of operations on portions of the Farm Welverdiend, water was sold to claimholders or diggers by the defendants and/or the Welverdiend Diamonds Ltd., and/or other companies herein before mentioned, commissions were received from diggers and prospectors, and diamonds recovered, including smaller stones known as "Bantams", and realised. Furthermore, rights to trade were sold on the Farm Welverdiend, either by the defendants and/or the Welverdiend Diamonds Ltd., and/or other companies herein before mentioned.

19. YOUR Petitioner attaches hereto copy of an extract from the books of the Welverdiend Diamonds Ltd., marked "F" which goes to shew that the working profit of the Welverdiend Diamonds Ltd. from 1st July 1927 to the 6th November 1927 was £142,164. 0. 9., that the working profit of Carlis in respect of the discoverer's claims from the 7th November 1927 to the 31st December 1927 was £8120. 8.8., and the working profit of J. Donaldson for the same period in respect of the owner's claims was £27,963. 9. 4., shewing a total working profit of £17$,247.18. 9. The extract shews that the profit for the period to the

30th June 1927 was £37,842. 10. 6., but from previous information placed before your Petitioner it would appear that for the period from the 1st February 1927 to the 30th June 1927 the proceeds from the sale of diamonds amounted to £44,408. 16. 3. against an expenditure of £10,312. 15. 3.

No discovery has been made of the books of the Welverdiend Diamonds Ltd., in relation to the recovery and realisation of diamonds by the Welverdiend Company, and Respondents, nor of the books and records in relation to the expenditure in regard to such realisation. From the fact that in the books of the Welverdiend Diamonds Ltd., there is reflected the working of the owner's and discoverer's claims by the Respondents personally from the 7th November, 1927, the probability, presumption or suspicion exists that in the working of the owner's and discoverer's claims from the date of purchase of portion P and others by the Welverdiend Company, that the Welverdiend Company was merely the nominee of the Respondents or that the Company was under the control and direction of the Respondents.

20. BY reason of the matters hereinbefore set out your Petitioner submits that he is entitled to further and better discovery in relation to the matters hereinbefore set out.

W H E R E F O R E your Petitioner humbly prays that it may please this Honourable Court to grant an Order that Defendants within a time to be determined by this Honourable Court, make further and better discovery of:-

1. The Certificates of all Discoverer's and Owner's claims now or heretofore held by the said Defendants or either of them.

2. Any contract of sale or other alienation of any Discoverer's or Owner's rights or claims to the Welverdiend Diamonds Ltd.

3. Applications and Declarations of Seller and Purchaser submitted by Messrs. J. A. Campbell and Marks for transfer of:

    200 Owner's claims on Portion E.

    200 Owner's claims on portion B.

    200 Owner's claims on Portion U.

    200 Owner's claims on portion P.

    200 Owner's claims on Portion K.

4. Correspondence between the Defendants, their attorneys or others on their behalf and the Honourable the Minister of Mines or his Department, on the subject, defendants' books and documents, Deed of Sale, Statement of Case and Counsel's opinion, and the Agreement with the Government referred to in the Honourable the Minister's statement in the House of Assembly on the 11th May 1928.

5. The Banking Account of the defendants or either of them kept at Lichtenburg or elsewhere into which the cheques received by the defendants or either of them from the Welverdiend Diamonds Ltd., were paid.

6. The books, records and documents of the Welverdiend Diamonds Ltd.,

7. Record of the number of diamonds extracted, the value thereof, and the proceeds realised therefrom by defendants or either of them since the coming into operation of Act 44 of 1927.,

8. All contracts entered into between the defendants or either of them and the Welverdiend Diamonds Ltd., or any other Company, corporation or person in relation to the sale, lease or other dealings of or with any Owner's or Discoverer's claims situate on any portion of the farm Welverdiend, the books and documents of the defendants or either of them relating to the ownership or control of any of such claims or the dealings therewith, the books and documents of the defendants shewing their operations on the claims, the precious stones recovered, the amount realised therefor or how otherwise dealt with, the expenditure on the operations, the books and documents shewing the returns to the Diamond Department and returns made to any Government Department in relation to the finds or disposal.

9. All Deeds of Sale, contracts, books and other records relating to the sale or other alienation or disposal of the Discoverer's and Owner's claims on any portion of the farm Welverdiend, the names of the purchasers and the consideration payable.

10. The deeds and documents of the sale of the shares in the R.U.V. Syndicate (Proprietary) Ltd., to Lichtenburg Gravels Ltd., shewing the consideration paid, the names of the shareholders who alienated their shares, and the names of the shareholders at the date hereof with the number of shares held by each shareholder.

11. The books of Account, contracts, banking account and all records of the Company, New Mines Ltd., in as far as such books, records or accounts reflect any operations on the farm Welverdiend, whether by defendants or others, and all moneys received from any Company carrying on or that has carried on operations on any portion of the farm Welverdiend or from the defendants or either of them.

12. The books, records and documents of the defendants and of New Mines Ltd., shewing all moneys received, whether by way of loan, dividends, profits, director's fees or otherwise from the Welverdiend Diamonds Ltd., and from any other Company that operated any claims on any portion of the farm Welverdiend.

13. All Balance Sheets and Profit & Loss Accounts rendered from time to time by the Welverdiend Diamonds Ltd., or other Company as aforesaid to the defendants or either of them or New Mines Ltd., as also of all reports made by the Welverdiend Diamonds Ltd., or any other Company as aforesaid to any Government Department of any Precious stones found, realisations and expenditure.

14. Records of all sales of water by defendants on the farm Welverdiend and/or any returns made to them or New Mines Ltd., by the Welverdiend Diamonds Ltd., or other Company as aforesaid of any such sales.

15. Records of all commissions received from diggers and prospectors by the defendants or either of them or New Mines Ltd., and/or any returns made by Welverdiend Diamonds Ltd., or any other Company as aforesaid to the Defendants or New Mines Ltd., of any commissions received.

16. Records of the sale of any Bantams recovered by the defendants from the farm aforesaid and/or of any returns made by them or New Mines Ltd., of any Bantams recovered by the Welverdiend Diamonds Ltd., or any other Company as aforesaid.

17. Records of the sales of any Trading Rights by the defendants or either of them and/or of any returns of such sales made to them or New Mines Ltd., by the Welverdiend Diamonds Ltd., or any other Company as aforesaid.

18. Books and records of the defendants disclosing the details and particulars of the items in the Account marked "A" attached to Defendants' Plea.

And for such further or other relief as this Honourable Court may deem meet, and that the defendants pay the costs of this application.

AND your Petitioner as in duty bound will ever pray.

BSchlesinger

I, BRUNO SCHLESINGER, of Johannesburg, the applicant in the above Petition, make oath and say that the statements and allegations contained and set out in the aforegoing Petition, are true and correct to the best of my knowledge and belief.

SWORN to before me at Johannesburg this 22nd day of November, 1928.

                                              Commissioner of Oaths

Printed in the United States
71210LV00001B